Prisoners, Lovers, and Spies

Prisoners, Lovers, & Spies

THE STORY OF INVISIBLE INK FROM HERODOTUS TO AL-QAEDA

KRISTIE MACRAKIS

Yale

UNIVERSITY PRESS

New Haven & London

Published with assistance from the foundation established in memory of Calvin Chapin of the Class of 1788, Yale College.

Yale University Press books may be purchased in quantity for educational, business, or promotional use. For information, please e-mail sales.press@yale.edu (U.S. office) or sales@yaleup.co.uk (U.K. office).

Set in Stempel Garamond and Futura types by Integrated Publishing Solutions.
Printed in the United States of America.

Library of Congress Cataloging-in-Publication Data

Macrakis, Kristie.
Prisoners, lovers, and spies : the story of invisible ink from Herodotus to al-Qaeda / Kristie Macrakis.
pages   cm
Includes bibliographical references and index.
ISBN 978-0-300-17925-5 (cloth : alk. paper) 1. Writing, Invisible—History. 2. Invisible inks—History. 3. Confidential communications—History. 4. Espionage—Equipment and supplies—History. I. Title.
Z104.5.M33 2014
652—dc23
2013032344

A catalogue record for this book is available from the British Library.

This paper meets the requirements of ANSI/NISO Z39.48-1992 (Permanence of Paper).

10 9 8 7 6 5 4 3 2 1

For
David Kahn

CONTENTS

PREFACE

This book grew out of my discovery of a top-secret invisible ink formula and method. Never before in the history of espionage had any government spy agency released or published a classified secret-writing formula and method.

It was the summer of 2006, and I was in Berlin. I had been submitting request after request to the Stasi archive for material relating to methods for secret writing, or SW, for my earlier book on methods used by the East German intelligence agency, commonly known as the Stasi. In return, I had been receiving file after file of detailed material on how to apply secret writing to a piece of paper and how to detect secret writing by the Stasi's mortal enemies, the CIA and the BND (West German Intelligence Agency), but despite frequent pleas, no secret formula or method for creating an effective secret ink had surfaced from the miles and miles of Stasi files. It was frustrating and disappointing. I was ready to give up.

Then on that summer day, the archivist handed me a thin file hidden underneath a pile of useless files. When I peered inside my mouth dropped open. Right before my eyes I saw a document stamped with the German equivalent of "Top Secret," and it had a formula written on the first page. My heart started pumping as if I were a kid who had just stolen a candy bar. My face flushed. I felt alive and awake. After skimming the file I knew enough to gauge its significance. My mind started racing:

where would I find a chemist to reproduce the reaction system? I began to transcribe surreptitiously the main two pages of the text. I could not believe the archive would let me make copies of the file or release it once they saw what was in it. After I finished copying the file by hand, I reread it so that everything would sink into my brain. At that point, I didn't know anything about cerium oxalate, the rare earth metal used as the secret-writing substance, but I quickly became familiar with it and the developing chemicals included in the file.

Even though I thought the archivists might have made a mistake, I put an order in to photocopy the whole file, along with a few other ones. I tried to be nonchalant about the request. As I was getting ready to leave the archive for the day, the staff associate, to my great surprise, handed me a copy of the file! As I leaped two steps at a time down the wooden stairs, I could hear my footsteps echo (or were they someone else's?) in the cavernous hall. I waltzed out of the building, looked behind me, saw no one chasing me, and fled the building. I finally had gotten what I was looking for.

The information I obtained would grow in significance when I paired it with the vague and impressionistic information I received about the CIA's methods. Because of a ridiculous secrecy policy, the CIA still refused to declassify World War I secret-ink methods and formulas, let alone modern ones.

When I returned to the United States later that summer, I sought out several chemists at my university to determine whether they would be interested in reproducing the secret-writing formula and method. One of them, my office neighbor, Dr. Ryan Sweeder, enthusiastically agreed, and we experimented, successfully reproduced, and published the secret formula with a couple of students in the lab prep room, later dubbed the "Spy Lab" by journalists.[1]

By this time I was wondering what else had been written about the subject of secret writing in its hidden or invisible form. The answer was: precious little. While bookshelves groaned under the weight of books on cryptography, the study of codes and ciphers, little had been written about its sister discipline, the history of hidden or covered writing, invisible ink, or the art of secret writing. This dearth of information became a problem when I scheduled a lecture and demonstration on the history of invisible ink for my class on the history of espionage and technology. The day before the planned lecture I wandered into the library and found numerous books on cryptography and cryptology, but not one single book on the history of invisible ink. Although I found a few references scattered in the standard cryptography history books by David Kahn and Simon Singh, from which I could cobble together a thin history, I spent most of the lecture producing simple demonstrations like the classic childhood lemon juice and heat experiment. At that point, I never imagined the rich stories I would find after digging into the dustbin of history.

My curiosity was piqued by the challenge of the hidden history. But it wasn't until years later, after I had completed *Seduced by Secrets,* and with the encouragement of David Kahn, that I decided to tackle this fascinating but hard-to-track history.

If the term "forensic traces historian" exists, I became one. The mantra of forensic scientists is "every contact leaves a trace." Just as the forensic scientist searches for traces of crime evidence, so too did I begin to search for trace evidence left behind by invisible-ink users and creators to write a nontechnical social and cultural history of invisible secret writing.

The main question animating my quest was: how important was invisible ink? In answering this question I was aided by the fact that British, American, and German archives had recently declassified or released exciting new material about spies and invisible ink from the two world wars and the Cold War that underscored its significance to spy agencies.

This book grew out of that discovery during a summer research trip to Berlin, was researched out of curiosity, and was written because of a need. It is the volume I wish I had found on the shelf.

■

Prisoners, Lovers, and Spies is a biography of a special kind of secret communication, the invisible kind. It tells the story of the life and times of secret writing from the ancient Greeks to the present day in its social, political, scientific, and cultural contexts; it is bookended on one end with the ancient Greek historian Herodotus's description of wax tablets and on the other end with al-Qaeda's hidden messages in digital pornographic movies. In between lie stories of international intrigue, of life and death, of love and war, and of magic and wonder. These tales about the people who used and created this secret communication method illustrate key turning points in its evolution through thousands of years of history. Spies were imprisoned or died, adultery was unmasked, and battles were lost because of faulty or intercepted secret communications, but successful hidden writing helped save lives, win battles, and ensure privacy; at least once, it even changed the course of history.

Unlike its big brother cryptography, the study of codes and ciphers, invisible secret writing didn't evolve in a progressive, linear fashion. It developed in fits and starts. While cryptography took off as science and was incorporated into government cipher offices during the Renaissance, invisible secret writing lagged in scientific sophistication until the early twentieth century. It finally caught up to, perhaps even surpassed, cryptography a century later with the emergence of digital image hiding on computers.

While cryptography early on experienced the perfect storm—advanced science and willing scientists, a volatile political context, and the personal experiences of the participants—invisible

secret writing had to wait until World War I for a similar con-
junction of circumstances to accelerate its progress.

This doesn't mean it wasn't useful. In fact, invisibility itself
made secret writing effective and sophistication unnecessary
during the early years; sometimes what's simplest is best. Sometimes the way you conceal secret writing is more important than
how you write it. While cryptography announces it has a secret
because it scrambles letters, unless someone suspects a message
is written in invisible ink, it usually remains a secret and doesn't
invite scrutiny.

In addition to illustrating critical points in the history of invisible ink, the stories of prisoners, lovers, spies, and scientists
also underscore the importance of invisible secret writing for
society and culture. Imagine a world without secret communication. The ability to send a message invisibly played a decisive
role in intelligence and counterintelligence, by keeping communication private, and even in entertaining and edifying us in
everyday life.

While most of invisible ink's drama played out in the world
of international intrigue, there was a long period when its leading role was in the world of magic and popular science. People
have been enchanted by the magical color changes of secret ink
for hundreds of years, but not many know what the missing
link is between Enlightenment popular science and our modern
fascination with magical disappearing inks.

For the reader wondering about terminology and definitions,
let me say a little about terms such as "invisible ink," "secret
ink," "secret writing," "invisible secret writing," and "steganography." I have used the phrase "invisible ink" in the title of
the book because most people will recognize it as meaning a
substance used to write invisibly, but my story is about secret
communication and hidden writing more generally—it includes
mail interception, invisible ink, microdots, and some digital

steganography. "Steganography" is an academic-sounding word that encompasses all the methods of hidden writing. Steganography is also a technical subject in computer science, but my book is not a technical treatise on modern steganography, nor is its focus on the twenty-first century. Steganography is not a dinosaur, although the word shares a root with "stegosaurus": the Greek word *steganos* means roof or cover. Just as a stegosaurus had spiky protrusions covering its body like armor, steganography is writing hidden under something else.

Although intelligence agencies born in the twentieth century use the term "secret writing," or SW, to refer to invisible ink and to shed the childish-sounding term, I didn't use that phrase in the title because cryptography, the study of codes and ciphers, is often referred to as secret writing. The proper phrase to describe the subject of the book is invisible secret writing.

Prisoners, Lovers, and Spies is organized chronologically, but readers can dip into periods they prefer. If the American Revolution is an interest, they will learn about how George Washington's "sympathetic stain" helped win the Revolutionary War. In the more fast-paced modern period, there are tales of German spies hiding invisible ink in a tooth or impregnating handkerchiefs only to be caught by the British through blanket mail interception.

The evolutionary thread and the impact of secret communication on history is the theme, but the book is also a narrative of stories linked in a broad contextual arc through thousands of years of history. Wars, political intrigue, society, scientific innovation and popularization, spy bureaucracies, and even gender all shaped the story in its transit through turbulent times. Finally, the book is selective, not encyclopedic. Readers wishing to know about every episode in this history are encouraged to consult the notes, archival sources, and Google Books.

The Art of Love and War

IN HIS RACY MANUAL ON seduction, *The Art of Love,* the Roman classical poet Ovid (43 BC–AD 18) introduced audiences to a very modern lover. He cruised the cobbled streets of ancient Rome, a large cosmopolitan city, to find the perfect pickup spot while handing out advice that could be recycled in Dear Abby columns or modern-day women's magazines.

The lover was a man about town. He strolled from grand arched piazzas to amphitheaters in his toga, he cheered at the Circus chariot races and lounged at public baths as he chased his quarry. Ovid's Rome was opulent, bursting with life and confident from the spoils of a conquered world. Disparaging his ancestors, Ovid declared that Rome had culture and refinement that included fine dress (casual chic worked best) and makeup (don't wear so much that it cakes on your face).

Written two thousand years ago at the dawn of the new millennium, the spirited, bawdy poem provided advice for men on how to find, catch, and keep a woman. Ovid wrote from firsthand experience. His wealthy father bankrolled his education in Rome, where he studied rhetoric, hoping to become a politician. After his education he traveled to Athens, Asia Minor, and Sicily. When he returned to Rome, he held minor positions as a government official. But Ovid grew bored, dropped out, and

1

began to write poetry for a living. Though not as symmetrically handsome as the sculpted busts of statesmen and gods the ancient Romans left behind—he had a long ski-slope nose and full, sensuous lips—he was kind and friendly and attracted women with his charm. A self-styled mediator of the battle of the sexes, Ovid also counseled women about the arts of allurement, drawing on the advice of Venus, the goddess of love. He even encouraged them to practice the art of deception, the deception of adultery.

Ovid taught young women, who were closely watched by their parents, how to hide a packet containing a secret message on the calf of a girlfriend or to stuff it in her bosom or to wedge it between the sole of her foot and her shoe. If this failed, he advised writing on the messenger friend's back. Since there was a chance that the male suitor might be attracted to the female messenger, Ovid warned the women not to choose pretty couriers. If the woman wished to bypass a messenger altogether, Ovid recommended writing a message with linseed oil on parchment.

But the most intriguing method Ovid suggested was using fresh milk as an invisible writing substance for secret letters. For centuries, this passage has been considered the earliest reference to a primitive form of secret ink:

> A letter too is safe and escapes the eye, when written in
>   new milk:
> touch it with coal-dust, and you will read.[1]

Even though Rome was comparable to late-twentieth-century America in its sexual freedom, Emperor Augustus banished Ovid from his homeland because of his raunchy poetry. It surely did not help in Augustus's moralistic eyes that Ovid had been married three times and kept a mistress during his second and third marriages.[2] There were also hints that he may have been complicit in an illicit affair conducted by Augustus's grand-daughter Julia. Ovid spent the last ten years of his life in exile on the Hellenistic backwater seaport of Tomi on the Black Sea (now Constanta, Romania) writing appeals for recall home in his *Sorrows and Laments*.[3]

Lovers have needed secret forms of communication since writing existed. They have used a variety of codes and secret writing to conceal their love letters and rendezvous from the rest of the world. Fear of ridicule or fear of discovery inspired lovers to communicate creatively and secretly. Secret communication is part of the game of love, and forbidden love heightens the romance.

But lovers were not the only ones who wished to express their thoughts in secret. People had found many ways to secretly communicate. Some wove secrets into a tapestry, while others told a tale to an apple and buried it in the earth. Ausonius, a Christian Latin poet who lived a few centuries after Ovid (c. AD 310–95), wrote lewd poetry in code and used milk on paper to conceal secrets to an old friend. He seems to have known countless codes for "concealing and unlocking secret messages," but he kept most of them secret from us as well. One of Ausonius's codes employed Greek letters to symbolize sexual positions.[4]

Apart from milk itself, another milky substance was used for invisible writing in ancient Rome. Pliny the Elder, the Roman naturalist, mentions in his vast encyclopedia of the world, *Natural History* (AD 77), the use of the tithymalus plant (modern-day spurge) for that purpose. Pliny reported having heard that adulterers traced letters on a body with the milk of the tithymalus plant, a medicinal herb also called goat's lettuce, and then allowed it to dry. When the body was sprinkled with ashes, the letters became visible. Ever the scientist, Pliny, whose studies were cut short when he was caught in the 79 eruption of Mount

Vesuvius, then launches into a discussion of the different kinds of tithymalus plants.[5]

The Art of Warfare

While the Romans used hidden writing in the art of love, the Greeks and Persians, several centuries before, had pioneered the use of hidden and invisible writing in the art of warfare. They especially honed their secret messaging skills during the Persian Wars, a series of conflicts between the Greek city-states and the ever-expanding Persian Empire during the fifth century before Christ (499–449 BC).

A remarkable episode from the Persian Wars is the tale of the man with the tattooed scalp. Histiaios, a Greek living at the court of the Persian king in Susa, the landlocked capital of Persia, needed to communicate with his loyal son-in-law, Aristagoras, on the coastal city of Miletus, Ionia, a Greek settlement of Ionians subjugated by the Persians. Histiaios wanted to exhort Aristagoras to lead a revolt, but spies and watchmen guarded the treacherous roads that wound through craggy mountains and cut a swath through wide open deserts. To make sure his seditious message remained secret, he summoned a trusted, illiterate slave to his quarters one day and shaved his head. Telling the slave that this would cure his bad eyes, Histiaios tattooed a message onto his scalp and waited for the hair to grow back. He then sent the foot messenger on his way to Miletus. When the slave messenger arrived at the door of Aristagoras, he shaved his head and displayed his scalp with its important message: "Histiaios to Aristagoras: make Ionia revolt." In addition to the weeks needed for the slave's hair to grow, the trip on foot to the coast must have taken at least three months, but what the plan lacked in urgency it made up in effectiveness: the revolt was a success.[6]

It is not only the modern reader who will be bemused by the

A drawing by Joy Schroeder evokes the
story of Histiaios's shaved-head message to
Aristagoras during the Persian Wars

amount of time it took the message to get to its recipient and
the peculiarity of the method. In the seventeenth century, John
Wilkins, author of *Mercury; or, the Secret and Swift Messenger,*
commented on the "strange shifts the ancients were put unto,
for want of skill."[7]

Dead animals were also used to send messages. When Astyages, king of the Medes (in modern-day Iran), discovered that
his relative Harpagos had not carried out orders to murder his
grandson and heir, Cyrus, he brutally killed, cooked, and dis-

membered Harpagos's son. To add insult to injury, Astyages surreptitiously served the remains to Harpagos on a silver platter at a dinner, revealing the menu only after the meal.[8] Since guards patrolled the roads, Harpagos slit open the belly of a hare in order to hide his secret message to Cyrus, the king of neighboring Persia, whom he had befriended.

After inserting a scroll in the dead hare's belly urging revolt against the Median Empire, Harpagos sewed up the incision and handed the hare to a trusted servant in a hunting net. The messenger ran to Persia disguised as a hunter with instructions to give Cyrus the hare and slit open the belly when no one else was present. Cyrus duly opened the belly and read the scroll. As a result of a secret message, cleverly concealed, Cyrus assumed the kingdom of the greatest monarchy in the world.[9] The ancients did not seem to mind the stench that the dead hare must have emitted by the time it reached its destination.

History is littered with similar examples of secret messages urging revolt. But one message in ancient Greece warning of surprise attack changed the course of history. Demaratus, a Spartan exile in Persia, watched as Xerxes, king of Persia, prepared a large army to invade Greece. Alarmed, Demaratus devised a way to hide a message using a wax-coated wooden tablet. Instead of writing on the wax layer, as would have been conventional, he scraped the wax away and scratched his warning into the wood. Then he melted the wax back onto the tablet, hiding the message. After the blank tablet reached Sparta, the king's daughter finally guessed the method of communication and scraped the wax off.[10]

With this advanced warning, the allied Greek city-states were able to prepare for the attack. Since the navy did not have enough boats to battle the Persians in the open sea, they lured the Persian ships into the harbor. While the Mediterranean blue

Battle of Salamis

sea still sparkled, the harbor was packed with the opposing triremes—boats outfitted with three layers of oars—but soon soldiers hurled their spears against the enemy in a sky filled with billowy clouds of smoke.[11] After the Greeks won the battle of Salamis, Xerxes fled to Persia and never attempted to invade Greece again. A hidden message on a tablet had kept the flame of Western civilization alive.

Sulla's Pig Bladder

By the first century before Christ, the battles in the ancient world had shifted from Greeks versus Persians to wars between the Roman Empire and anyone who resisted its imperialist aims.

One of Rome's deadliest enemies turned out to be the "Poison King," Mithradates, who ruled Pontus, a small kingdom on the Black Sea, in what is now northeastern Turkey. He was so disliked in the Western world that his enemies exaggerated his thick features—a broad nose and thick brute lips—to emphasize his Oriental, despotic nature. Mithradates was hailed as a military genius, and his knowledge of poisons allowed him to crush enemies, foil assassination attempts, and get rid of rivals. He imagined an eastern empire that would surpass the Roman one. His first step toward conquest was to exterminate Roman citizens living in Anatolia (modern Turkey) and the Aegean islands, Rome's newly conquered territories. In the spring of 88 BC he masterminded a plot to incite local leaders in dozens of cities "to kill every Roman man, woman, and child."[12]

It remains one of those historical mysteries how Mithradates managed to keep the plot secret from the Romans, since so many local leaders knew about it. There is no doubt that he communicated secretly with his coconspirators to plan the massacre. On the appointed day, the natives rounded up and killed all Romans and Italians living in their towns. At least 80,000—some say 150,000—Italian residents of Anatolia and the Aegean islands were killed. To avenge this shocking bloodbath, the Senate sent Rome's most celebrated general, Lucius Cornelius Sulla, to battle Mithradates.

If Mithradates' deeds were scary, Sulla's appearance and character were terrifying. He was born into a poor though patrician family and found a patron to finance his political career and insatiable appetite for power. As a military leader he had a commanding presence and was considered a crafty fox and a brave lion. Although he had regular features, golden-blond hair, and sparkling blue eyes, his skin was very fair and turned blotchy with red patches when he was older. People joked that Sulla's face looked like a pizza, with "purplish-red mulberry mash sprinkled with white flour." His complexion, along with his arrogant personality and piercing eyes, was intimidating.[13]

Sulla's first step toward avenging Mithradates' massacre was to occupy Greece. In his battle for Piraeus, Athens's fortified port, Sulla collected tens of thousands of mules to help operate his siege engines and towers. To build the engines he decimated all the nearby olive groves, held sacred in the Greeks' religion. For whatever reason, two men inside the Piraeus walls decided to betray Mithradates' campaign and warn Sulla of his plans. They inscribed secret messages onto lead sling balls and threw them over the fortress walls. After the balls began to collect, Sulla grew curious and picked one up. He read that Mithradates' forces were going to attack workers, and the cavalry was going to charge his army. With this warning, Sulla was able to ambush and kill the enemy troops before they assaulted his own.[14]

As siege warfare continued, the traitors within Mithradates' camp continued to volley more balls over the fortress wall. This time the messages described secret night shipments of wheat to Athens. Sulla was able to ambush further food supplies to Athens, and pretty soon the city was starving.

Mithradates' army dwarfed that of Sulla—his army from many different lands numbered 120,000 barbarians, while Sulla had only about 30,000 troops. But Sulla had more spies. And he learned how to communicate with them secretly by reading a manual for defense against siege warfare written by the general Aeneas Tacticus, or the Tactician, in the fourth century BC. Sulla especially liked inflating a pig's bladder so that it resembled a modern-day football (itself once called the pigskin) and writing on it with a mixture of ink and glue. Once the writing was dry, the air was let out of it and the bladder forced into a glass flask. The flask was filled with oil and corked. Once the recipient had the flask, he poured out the oil, inflated the bladder, and read the message. Then the spy replied the same way.[15]

Through secret methods like these, Sulla's spies informed him that Mithradates' army had moved southeast and camped in rocky hills. Sulla's spies knew a secret trail high above Mithradates' camp and proposed showering the army with huge boulders, forcing them onto the plain below. The surprise attack worked and Sulla was able to kill thousands of enemy soldiers, leading to victory in Greece.[16]

The Tactician

Sulla was not the only general to learn about secret communication methods from the prominent but faceless Aeneas Tacticus, who wrote the first manual for defense against siege warfare. An adventurous mercenary soldier, Aeneas liked to travel and was well read. *How to Survive under Siege,* widely consulted in ancient times, remains a classic on the art of warfare. Addressing such topics as effective guard duty, signals, gatekeepers, scouts, and censors, Aeneas focused on countering traitors from within. His work was so influential that one of his secret-writing methods was still used more than two thousand years later in the twentieth century: the sender of a secret message lightly dotted selected letters in the first three lines of a book or in a newspaper. The recipient then rearranged the selected letters to decode the message.[17]

In addition to writing on pigs' bladders, ancient messengers transported secret messages written on leaves plastered to a leg, women wore fake earrings made of thin pieces of rolled-up beaten lead bearing inscriptions, a traitor sewed a message of betrayal on papyrus sewn under the flaps of his breastplate, and another man sewed a sheet of papyrus to the bridle rein.[18] When human messengers were not fit for the situation, spies sent messages hidden in dog collars.

So far, Greek and Roman warriors were really using "hidden

writing," not invisible ink. But Aeneas mentions ink as the writing medium in a method that is a modification and extension of the tablet method used during the Persian Wars. Demaratus engraved the message into the wood, but Aeneas recommended writing on a boxwood tablet with the best-quality ink, letting it dry, then whitening it to make the letters invisible. The whitening substance could be a variation of the gypsum Greeks used to whitewash their houses, or white slip, an easily removable clay substance. When the tablet arrived, the recipient placed it in water, washing away the white substance, and the letters appeared.[19]

Let There Be Gall Enough in Thy Ink

For centuries children have used lemon juice and heat as a simple method of invisible writing and revelation. Especially given the ubiquity of the lemon in Mediterranean climates, and the part the fruit plays in the region's cuisines, we might expect that primitive form of secret communication to have been a forerunner to more sophisticated methods. But I have found no evidence to document the use of lemon juice for invisible ink in classical times. Apparently, no one happened to wave a lemon juice–infused piece of papyrus over a campfire to discover the magical qualities of lemons.

And while Aeneas recommended more than a dozen techniques for hidden writing, he never revealed what kind of ink he used, nor did he demonstrate knowledge of the concept of a reagent—a second fluid to make the invisible visible. For this moment in history we have to wait one more century for the work of Philo of Byzantium, who flourished c. 280–220 BC in Alexandria. A brilliant Greek engineer, Philo shared Aeneas's interest in siege warfare, but his passion was the mechanics of attacking fortresses using catapults. Philo also provided advice on sending a negotiator into a besieged town using secret writing: "One writes this letter on a new hat or on human skin with crushed gallnuts dissolved in water. When the writing is dry it will become invisible. Then soak a sponge with vitriol [ferrous sulfate] and rub it over the invisible writing and it appears."[20]

Philo's advice is the first account of a sophisticated invisible ink method using a reagent. For the modern reader, gallnuts seem like a foreign and uncommon substance, but in the ancient Mediterranean world they were quite common. Not really nuts, gallnuts, or nutgalls, are nutlike swellings produced on an oak tree by parasites or insects like wasps, which lay eggs on the branches. Most of the liquid produced in the swelling, or tumor, is tannic acid or tanno-gallic acid, a brownish or yellowish substance used in tanning and dying, or as an astringent.

For thousands of years, gallnuts provided the main ingredient for most regular black writing inks; they were also used to dye hair black. Gallnuts were abundant in Persia, Mesopotamia, Syria, and Asia Minor, although the rarer Aleppo gall from Syria remained the most highly prized of this sort of ink for thousands of years. Mixing a solution of galls with gum Arabic, water, and ferrous sulfate, a salt of iron known as copperas or green vitriol, produced an inky, indelible, black substance.[21]

Iron-gall ink was the most important ink in Western civilization. Bach composed with it, van Gogh drew with it, and prolific authors from Leonardo to Shakespeare wrote with it. Ferrous sulfate is as green-blue as the sea and as caustic as acid in its crystal form, so the term "green vitriol" seems quite appropriate; when Shakespeare's Sir Toby Belch, in *Twelfth Night*, challenges Sir Andrew to have "gall enough in [his] ink," he is demanding boldness of action, not just bold writing on the page.[22]

Users of ink made from galls and iron sulfate soon discovered that a message written with one ingredient could not be seen, but brushing over it with the second ingredient made it appear black just as if the two had been mixed from the start. That Philo of Byzantium described such a sophisticated invisible ink formula in the third century before Christ—without even a rudimentary first step like lemon juice—is not only remarkable but also a mystery. In contrast to organic invisible inks like lemon juice, milk, urine, and saliva, each of which can be developed using heat, gallnuts need a specific developer—a reagent—to become visible. Did the Greeks really already develop one of the most sophisticated methods for secret writing requiring a reagent thousands of years ago? Or did later commentators add the chemical specificity in the passage quoted above? Whatever the case, as we shall see in later chapters, gallnuts and iron sulfate played an important but mysterious role in the American Revolution.

From Rectal Concealments to Urine Ink in the East

While the ancient Greeks carved messages on wood tablets and then dipped them in wax, the Chinese wrote secret messages on silk or paper, then rolled it into a ball—*le wan*—and covered it with wax. The emperor or general then gave the wax ball to a messenger, who swallowed it or placed it in his rectum. Even though a stomach or rectum is a better hiding place than a blank tablet, retrieving a message was a little messier in China than in Greece or Rome.[23]

More than a dozen centuries later, secret-ink messages were considered quite magical in China, but like the Greeks, the Chinese used whatever they found in their environment to prepare messages. They used alum (aluminum potassium sulfate), the white substance used in styptic pencils, dyeing, and pickling. An invisible message written with alum appears when dipped in water. The Chinese, who mined alum, were using alum as invisible ink and developing it with gallnuts—spread across the globe by trade—by AD 980. In the twelfth century, the son of a military man named Wang Shu spread rumors about Ch'in Kuei (1090–1155), a leading Chinese government official during the early Southern Sung dynasty. Wang was stripped of his title and expelled. During his exile, Wang's son met a magician who showed him how to write invisible characters on paper that would reappear when dipped in water. Wang's son, still seething with anger, was curious about this magical ink. He wrote four invisible characters, "death to Ch'in Kuei," on a piece of paper. The magician was going to turn him in to the authorities, but a sizable bribe kept him quiet.[24]

Those looking for lemons in the early history of secret writing will eventually find them in an Islamic contribution to the science, though not in the simple technology employed by children. The Greeks had characterized Persians as barbaric and primitive, but by the eighth century, Islamic civilization had surpassed Greece in splendor. The Persian Empire had expanded into a broader Arabic-Islamic one that included the lower part of the Mediterranean basin and the Middle East. Along with territorial expansion came a flowering of culture and science. The rulers of the Islamic Caliphate opened the House of Wisdom, a translating school, library, and research center, in Baghdad and started translating the world's knowledge into Arabic. Not just copiers, the scholarly Arabs made their own valuable contributions, especially in mathematics and chemistry. Along with algebra and statistics, "cryptology was born among the Arabs."[25]

The new Islamic state spread the study of Arabic through the Koran, developed a hefty bureaucracy, and hired secretaries, tax collectors, and managers to run affairs of state.[26] *Secret*aries became synonymous with secrets. In the Arab-Islamic world the spy wasn't the bearer of secrets; instead, secretaries were considered the most important government officials and were entrusted with secrets of states that needed to be communicated through the mail. Protecting sensitive state secrets was a driving force in Islam for concealing messages: the Arabs pioneered cryptography, created cryptanalysis by using statistics and letter frequency, and practiced the art of secret writing.

The Arabs also needed to conceal messages sent between secretaries in Eastern and Western lands. Qalqashandī (1355–1418), a secretary in the chancellery in Cairo, in 1412 completed a manual for secretaries—*Subh al-Asha*—that provided an overview of existing cryptographic methods. Before sending an encrypted message the sender wrote with something invisible on paper. The sender and addressee generally had an agreed upon substance to make the message visible; otherwise, the receiver was advised to brush something on it or to hold it over fire. In addition to the known method, using fresh milk (this recipe added sal ammoniac), the Arabs started using onions crushed in water as an invisible substance that could be developed by heat. Like the Chinese, they used alum developed by water. Reversing Philo of Byzantium's method, they used copper sulfate (known as blue vitriol) as the invisible writing and brushed it over with oak galls crushed in water as a developer. By the thirteenth century they also used the milky substance from wild figs for invisible writing that turned red when heated.[27]

But the most remarkable process used in the Arab world involved a secret recipe using the lemon. The recipe calls for frying an equal amount of black (dried) lemon with colocynth root (a strong laxative sometimes called bitter apple or bitter cucumber) in olive oil and then mashing the mixture together. Egg yolk paste is mixed into the mash to make an invisible ink that can be used on any object. After a while, hair was supposed to grow where the writing had been—and not only on a head—making the text legible. As with the slave courier of Histiaios, it took some time for fifteenth-century Arab messengers carrying letters through dry deserts under the hot sun to reach their destination.[28] Whether the recipe worked or not, the Arabs seem to have been the first people to have used lemons for writing secret messages.

It is not surprising that the Arabs used lemons as a medium for hidden writing, since lemon trees and paper were readily available in the Arab world. No one knows exactly where lemons originated, but there is no doubt that lemon trees and bushes came from tropical and subtropical southeast Asia. While India and China possessed wild-growing lemon bushes, the Arabs cultivated lemon trees and brought them to the western Mediterranean, especially the Iberian Peninsula and Italy. Using lemon juice as invisible ink became quite popular in sixteenth-century Italy, where lemon trees had been cultivated for a century.[29]

The migration and diffusion of lemons among different civilizations and cultures in the Mediterranean during the early medieval period coincides with the rise of occultism and magic. This heady mix stimulated the use of invisible ink. When we think of occultism today, it conjures up images of Ouija boards and séances. But "occult" originally simply meant secret or hidden, evolving later to suggest hidden wisdom; thus secret writing was sometimes called occult writing. Practitioners of the hermetic arts that involved alchemy and magic disguised their secret knowledge with occult writing to keep it from the prying eyes of the vulgar.

By the thirteenth century secret writing also began to be practiced in the monasteries and employed by artisans to protect their trade secrets. Later, some secrets were compiled and bound into manuscripts called books of secrets. Everyone wants to know a secret. Therefore it's not surprising that after the introduction of the printing press in 1440, these books became best-sellers. During the early medieval years the secrets contained in these volumes described lost esoteric wisdom, forbidden knowledge, and artisans' secrets. By the Renaissance, the books of secrets became collections of recipes or techniques on how to make things like dyes, pigments, or even inks, alchemical procedures for making metals, household advice, and even cooking recipes. They were more like modern-day cookbooks or how-to books than books revealing the secrets of nature.[30]

Despite the seriousness surrounding secret writing, medieval people began also to notice its magical qualities. They even used such arcana in parlor games. But the occult and religious symbolism were ever-present. In a recipe for invisible writing using dust and urine, the word "urine" is coded: "Take xskova [urine] and write with it on your hand and when it dries nothing will appear at all. When dust or ashes are sprinkled on the invisible urine, it appears readable. If you want to play a trick with your friends, write 'no' on your hand with the urine and let it dry. When you want to joke around 'to know whether someone is a virgin' you sprinkle dust on the word no and then draw a cross on the dust. When your friends aren't looking, you brush off the dust and the word 'no' will appear." The medieval magician concluded that "Many onlookers will consider this a magic art . . . brought about by the sign of the cross."[31]

During the medieval period, between the fall of the Roman Empire and the early Renaissance, such organic substances as urine, lemon juice, figs, and onions began to be used widely for invisible ink. Beginning in the early Renaissance, knowledge about secret writing hidden in the monasteries, in the scholar's study, and in the craftsman's workshop became more publicly available with the invention of the printing press and the dissemination of knowledge to a wider public.

∎

Invisible ink arose out of the ancient arts of love and warfare. In both realms, fear was the mother of invention. These needs have remained remarkably constant through thousands of years since. Over the centuries invisible ink awoke from its dormant state in times of war; in times of peace, magicians and lovers played tricks or exchanged billets-doux. These ancient imprints may seem primitive to the modern world, but the principles and the human nature behind them have set the standard for the modern era.

Intrigue and Inquisition

GIAMBATTISTA DELLA PORTA's interrogation by the Roman Inquisition was the most terrifying event in his life, yet he never talked about it. Mysterious and secretive, della Porta was one of Renaissance Italy's greatest scientists and polymaths, full of youthful enthusiasm for the wonders of nature.[1] He was also the first to publish a major book on cryptography and invisible ink; he thought that invisible writing should be "faithfully concealed" for "great men" and "princes," yet he said, "This I will publish."[2]

Although della Porta was a scion of a noble family from Naples, not much is known about his personal life except that he adored his two brothers and that a mysterious woman—perhaps his wife—bore him a daughter who in turn produced two grandsons. Educated by private tutors, della Porta and his two surviving brothers never received university degrees but were groomed as courtiers and wrote essays in Latin and Italian by the time they were ten years old. Della Porta's boundless enthusiasm for the wonders of nature permeated his work, which spanned the fields of mathematics, optics, alchemy, astrology, physiognomy, memory, agriculture, and cryptography, to name the most important. A Renaissance man in every sense of the word, he also enriched Neapolitan cultural life with seventeen noteworthy plays.

ASPICIT ET INSPICIT

IO·BAPT·PORTAE
NEAPOLITANI

MAGIAE NATVRALIS
LIBRI XX.

Ab ipso authore expurgati, & superaucti, in quibus scientiarum Naturalium diuitiæ, & delitiæ demonstrantur

CVM PRIVILEGIO.
NEAPOLI, Apud Horatium Saluianum. D. D. LXXXVIIII.

Pompeij Sernij

Frontispiece, Giambattista della Porta, *Natural Magic*

Della Porta was also a magus, a wonderworker who had penetrated the secrets of nature. He sometimes even looked like a magician. A man of medium build, he had a bald sloping forehead, a long crooked nose, dark hair with a well-kept beard, and playful, bright eyes. He wore a black cape and ruffed collar typical of the Elizabethan age and often an accompanying black hat. Some engravings in the German adaptations of his international best-seller, *Natural Magic*, depict him with a sorcerer's wand and magician's hat.

Della Porta's life and work spanned the heart of the sixteenth-century Renaissance, coincided with the flowering of scientific societies, and met with the establishment of the Inquisition in Rome. More than a work of popular science, *Natural Magic* presents "the riches and delights of the natural sciences." The four books published in 1558 contained wonders of the world, but also the quotidian, from curiosities like monstrous births to descriptions of animals worthy of Pliny, from sketches of the properties of exotic plants to simple housekeeping hints. By the time of the greatly expanded second edition in 1589—an additional sixteen books—della Porta included a new chapter on the wonders and delights of "invisible writing," producing one of the most complete overviews ever published.

Invisible writing and cryptography appealed to della Porta's love of secrecy. His influential and encyclopedic book on Cryptography (*De furtivis literarum notis,* 1563), available only in Latin, was the first published work of its kind and is valued by all historians of cryptography, making him the "most outstanding cryptographer of the Renaissance," according to David Kahn.[3] Although he is better known for his contribution to cryptography, including the digraphic substitution cipher (a pair substitution), invisible writing intrigued him. He didn't make any original contributions to the art of hidden writing; rather, he described what existed at the time.

Giambattista della Porta dressed as a magician

Like other Renaissance humanists, della Porta thought his modern age was more sophisticated than the ancient world. While the ancients had used hidden or concealed writing, he opined that "our age has cleverly contrived invisible methods of writing in place of the former." Invisible writing was advantageous because the recipient had to know in advance to expect it, or it would "escape the sight of the beholder."

Invisible writing was even more effective than using ciphers alone because unless a potential interceptor knew to test for secret writing, he would never even suspect the existence of a message. Della Porta thought that invisible writing, especially when it was combined with some sort of cipher, offered "great obscurity in affairs of the greatest moment, when concealment is no small advantage."[4]

Around 1560, roughly when della Porta published his first two books, he hosted a group of scientists at his house to conduct experiments and discuss the wonders of nature. This group soon turned into one of the world's first scientific societies, the Accademia dei Secreti, who also called themselves the Otiosi, or men of leisure. No one was admitted to the new academy unless "he had discovered some new secret of nature useful in medicine or the mechanical arts."[5]

The fruits of the academy experiments and discussions were incorporated into the second edition of *Natural Magic.* Della Porta tells us that he "never wanted . . . at my House [for] an Academy of curious Men . . . [who] cheerfully disbursed their Moneys . . . and assisted me to Compile and Enlarge this Volume." In the intervening years, he also went on the grand tour of Italy, Spain, and France, visiting scholars and libraries and collecting information. Much of his knowledge and many of the examples used in his invisible-writing chapter came from his travels and from the academy members. He even gave Philip II, the king of Spain, a copy of his cipher book.[6]

But the academy did not last long. Della Porta was quickly denounced to the Inquisition by Neapolitan neighbors and summoned to the Holy Office in Rome to explain reports of witches' salves and necromantic arts. Not only was the name of the scientific academy suspect, but della Porta was dubbed "the Magus" in public and accused of dabbling in magic.[7]

Della Porta's brush with the Inquisition was more serious than the whitewashing attempts of friends and family would imply. They tried to save his reputation by circulating the rumor that he had been so persuasive in showing that his secrets were natural that he had received a commendation. In fact, he was summoned to the Holy Offices at least twice between 1574 and 1580, and his books had to be vetted thereafter before publication. As a result of the first summons, he was forced to dissolve the academy. When he was brought before the Neapolitan Inquisition in 1580, he was accused of having written about wonders and secrets of nature.[8] He was ultimately acquitted by the pope, but was warned to refrain from engaging in illicit arts. He joined the Jesuit order and outwardly conformed.

Della Porta never met with the horrors of Giordano Bruno, who was burned at the stake, or Thomas Campanella, imprisoned for twenty-seven years. If anything, the encounter with the Inquisition made della Porta and his circle even more secretive. They communicated in cipher and invisible ink.

Although della Porta thought invisible writing should be "faithfully concealed" for "great men" and "princes," he was the first to publish simple but diverse recipes for public consumption, including lemon and onion juice revealed by heat, glutinous substances by dust, and substances like alum by water or gallnuts. One of the most striking recipes recommended trapping a dormouse and extracting its juices. They glowed with a "fiery color" at night.[9]

By della Porta's time, using lemon juice or any fruit juice to hide secrets was well known. It was also known that the invisible writing became "visible through fire." Even though lemon juice as a secret writing substance had originated with the Arab world around AD 600, it was not widely used until the sixteenth century in Europe, and by 1558 contemporaries observed that "lots of people use lemon and onion juice that can be read by the fire."[10]

A smaller circle of people also knew that many common substances could be made visible by dipping them in water or rubbing dust on them. As we have seen, the ancient Romans were familiar with using the milk of spurge (tithymalus) to write messages on human bodies and papers, and the Arabs had discovered fig tree sap as an invisible substance. Both of these sticky or glutinous substances could be developed by dust or coal dust. During della Porta's time substances other than the juices came to be used more commonly. The most prominent of these invisible-writing substances was the gallnut, the growth on oak trees mentioned by Philo of Byzantium.

Della Porta's immensely popular *Natural Magic* includes advice on how to smuggle invisible ink–bearing eggs into prison because, he claimed, eggs were "not stopped by the Papal Inquisition." He almost becomes obsessed with this egg: "An especially safe method is to write letters on an egg, which will not fail to perform its errand of deception." Of course, once della Porta revealed the secret it may have become impossible to "perform" an "errand of deception"![11]

Even though della Porta was enthusiastic about using an egg for secret writing, it has been very hard to get this kitchen experiment to work. He recommends using alum and vinegar as the invisible substance with which you write on the egg. This secret egg experiment has been handed down in history for hundreds of years. Supposedly French peasants used it during World War I. In 1965 the *New York Times* announced that the Department of Agriculture was encouraging children to write secret messages on eggs with alum and vinegar in order to entice them to eat more eggs.[12] As late as 2012, the secret-writing egg was still frequently featured on the Internet. When Jason Lye, a color chemist, and I wrote to ask why our experiment didn't work, we got no reply. I've tried this experiment numerous times with my students and in preparation for demonstrations at talks. It doesn't work! I've tried writing on the egg with store-bought alum and vinegar using both a boiled and a raw egg. It never works. It's hard to believe that this recipe has been handed down for almost five hundred years and it doesn't work. We have failed to perform an "errand of deception" with a secret-writing egg.

Della Porta advised writing on an egg because the Inquisition wouldn't find it suspicious, but individuals who clashed with church or state weren't the only ones interested in developing invisible ink. The rise of modern diplomacy in Italy led to a practical need for coded messages, and offices employing cipher secretaries sprang up in the city-states. In this atmosphere of intrigue and secrecy new ciphers proliferated. Cryptology and secret writing flourished. Cipher secretaries and manuals on cryptography proliferated.

It is less commonly known that the state was also interested in procuring new invisible-ink formulas. As early as 1525, the Council of Ten, the Republic of Venice's secretive governing body and overseers of a secret police force, paid Marco Raphael, a leading cipher secretary and renegade Jewish convert to Christianity, 110 ducats—about the equivalent of what a cipher secretary earned in a year—for new invisible-ink recipes.[13] Raphael's invisible-ink characters could be made visible by placing them over heat, dipping them in water, or rubbing them with burned paper or gum or a metallic substance. Thus they are comparable to the recipes della Porta described later in the century. By 1531 Raphael had left for London and gained favor with Henry VIII by supporting his divorce. It is likely that he brought his secret invisible-ink recipes with him to England.[14]

A few years after Raphael sold his invisible-ink recipes, a small book appeared called the *Secreti* (c. 1530). It was printed with blank pages that could be dipped in water or held over fire to reveal invisible ink. The author, G. B. Verini, a poet, mathematician, and calligrapher, called himself a professor of writing. Opposite the first page of arabesque roman capitals, A & B, is a blank cartouche with directions at the top inviting the reader to dip the page in clear water to reveal a set of verses. The alphabet continues two letters at a time until R and S. Each page has another cartouche, a set of verses, and another recipe for secret writing. The required ingredients and tools include lemon juice, charcoal, a mirror for reversed writing, and a bone. Though Verini has not been included in the professors of secrets tradition, he certainly falls under that category. His book was probably peddled at bookstalls in the squares of Venice, along with other books of secrets containing practical how-to recipes.[15]

Elizabethan Espionage: Off with Her Head!

Della Porta's books *Natural Magic* and *De furtivis* were widely read on the Continent and in England. Scholars, scientists, and diplomats like Sir Francis Walsingham (1530?–90) who made the European tour might have encountered della Porta at the baths. Walsingham certainly learned about the work of della Porta's predecessors Leon Battista Alberti and Girolamo Cardan. The son of a prominent London lawyer, Walsingham studied at King's College, Cambridge, but as a Protestant felt forced to leave England with the accession of the Catholic Queen Mary I. He fled to Padua, one of the most picturesque towns in the

world, to study law at the university (1555–56). During his exile
he soaked up the culture and politics of the Italian Renaissance.
Luminaries like Galileo, Vesalius, and Copernicus had taught at
the university, and the spirits of Leonardo and Machiavelli still
influenced the culture. As a Protestant exile on the Continent,
Walsingham also lived in Switzerland and Strasbourg before returning to England when Elizabeth I became queen in 1558.[16]

Walsingham brought back Alberti's manual on cryptography and the Cardan Grill, a transposition cipher, along with a
deep understanding of Italian intelligence methods.[17] Elizabeth
learned about Walsingham through her chief adviser and secretary of state, Sir William Cecil, later Lord Burghley, who had
been in charge of intelligence work and had taken Walsingham
under his wing. As Cecil spent more time leading the government, he delegated secret work to Walsingham. After a stint as
English ambassador to France, Walsingham succeeded Cecil
as secretary of state in 1573 and remained in the post until his
death in 1590.

Walsingham was good at keeping secrets. He listened and
watched, staying silent until he pounced on his prey. He quickly
became Queen Elizabeth's spymaster, one of the first the world
had ever known. Furtive and satanic-looking, Walsingham usually
dressed in black clothes in a Technicolor Elizabethan Renaissance world of purple velvets, canary yellows, and festive reds.
The queen called him her "Moor" because of his dark complexion, pointed chin, and black hair and beard. His white Elizabethan
collar served to accentuate the blackness. Even though no formal modern spy agency existed in Elizabethan times, Walsingham has been considered the father of modern espionage. But he
was a one-man show. He even financed his own operations and
died in debt.[18]

Queen Elizabeth's interest in, and appreciation of, espionage
is reflected in the resplendent dress she wears in a famous por-

Portrait of Sir Francis Walsingham,
attributed to John De Critz the Elder, oil on
panel, ca. 1585

trait by Isaac Oliver. Elizabeth holds a celestial rainbow, and on
her sleeve is an orb-crowned serpent, symbolizing the marriage
of wisdom and power. But the dress is also covered with eyes
and ears, emblematic of the state's effectiveness in intelligence
gathering. Walsingham's spies were her ears and eyes domes-
tically and abroad. The metaphor is an apt and enduring one;
throughout history, intelligence services have aspired to develop
all-seeing, all-hearing spies.

Reportedly, Walsingham had an army of fifty-three spies
scattered on the Continent, with the lion's share in cities in
France and Spain, and some in Flanders and Holland.[19] They
were a motley crew of merchants, travelers, foreign exiles, cryp-

Portrait of Elizabeth I (1533–1603), Isaac
Oliver (ca. 1565–1617), color lithograph

tologists, former prisoners, writers (maybe even the playwright
Christopher Marlowe), and Catholic recusants who opportu-
nistically shifted their loyalties to Protestantism.

He recruited agents and informants for the two main threats
facing Elizabethan England. At first, the Puritan Walsingham
concentrated on domestic security and counterintelligence,
working to foil Catholic plots to kill Elizabeth or to mount an
invasion of Protestant England by Catholic exiles on the Continent. Later, he turned to foreign intelligence to prevent a sea invasion of the British Isles by the Spanish Armada. But his main
job was to keep Elizabeth alive.

One of Walsingham's most celebrated successes was catching
Mary, Queen of Scots, in the act of plotting with her support-

ers to overthrow Queen Elizabeth. Elizabeth had placed Mary
under house arrest out of fear that she might try to usurp the
English throne and install Catholics in the government. Mary
remained imprisoned in various castles and manors for nineteen
years. As a royal prisoner she had an entourage of attendants,
including a laundress, maids, cooks, a coachman, and even a cipher secretary. During the early years of her captivity she could
leave the castle for the countryside, where she would talk to
people and distribute alms to the poor, but later Elizabeth cut
off Mary's contact with the outside world.[20] Soon Mary was
unable to communicate with her friends and supporters on the
Continent and in Scotland.

At first, with the Earl of Shrewsbury as her jailer, Mary had
an easier time communicating with the outside world. She wrote
her letters in cipher and invisible ink and hid them in wine bottles or slippers or even a looking glass. But Walsingham tightened her communication channels one by one as he uncovered
plot after plot to murder Elizabeth. A plot by the Catholic Ve-
netian banker Roberto Ridolfi in 1570 to assassinate the queen
underlined the dangers of allowing Mary to communicate with
her friends on the outside. Now all of her correspondence had
to be handed to her jailer Shrewsbury, who passed the material
on to Walsingham and his chief cryptographer Thomas Phelippes. But this did not prevent her from creating other networks
of communication.

When an eighteen-year-old Scottish boy was arrested in Scotland in 1574, he confessed that he worked as a courier for Mary,
delivering mail to Scotland. A London bookseller had passed
packets of letters to and from a tutor to the children of Shrewsbury named Alexander Hamilton, and these were then handed
to the boy. The conspirators were brought to the Tower of London, but Elizabeth did not want to pursue the case further and
it was dropped.[21]

When Mary's other conspirators heard about the close call, many of them fled to Paris. One of these supporters was Thomas Morgan, who had once been a servant for Shrewsbury. In Paris he orchestrated a huge effort on Mary's behalf by organizing all her supporters and friends. Walsingham, though, continued to penetrate Mary's correspondence and was remarkably adept at turning caught Catholics into double agents.

While in Paris, Morgan placed himself at the service of the Bishop of Glasgow, who conferred with Mary on matters of strategy. Moreover, Mary was becoming aware that her communication channels might be compromised. In 1577 she wrote to the bishop with advice: should "access to me and news [be] restricted," she recommended that he "write in white [invisible ink] with interlines (alum seems to be the best or nut-gall)," under the pretense of sending a book. And although "such artifices," she continued, "be very hazardous and vulgar, they will serve me in extreme necessity by way and conduct of the carrier of this place, who is not so closely observed but that among the other necessaries which he brings me, he can deliver to me safely that which one will write to me in this manner, without perceiving it himself."[22] Mary and the bishop conversed freely in their correspondence, not knowing that Walsingham had intercepted those letters also.

Michel de Castelnau, also known as Monsieur de Mauvissière, the French ambassador to England, was another friend of Mary's. Castelnau was everyone's friend. He had tried unsuccessfully to achieve reconciliation between Mary and Elizabeth. Mary began to rely on Castelnau to send secret messages to her agents in Paris. She instructed Castelnau in a letter how to use alum as invisible ink: "The best . . . secret writing is allum soaked in a little clear water twenty-four hours before one wishes to write, and to read it, it is only necessary to damp the paper in some basin of clear water." The secret writing appeared white and could be read until the paper dried again. She recommended that he write on white cloth or fine linen and cut a piece at the corner as a signal that something was written on it. She advised using this method of secret writing only "on occasion of great importance."[23]

But even this correspondence could not be kept secret from Walsingham's prying eyes. He had received information in 1582 that the French ambassador in London had been acting as an agent in a new secret courier system between the imprisoned Mary and her supporters in France.

Walsingham was a longtime friend of Castelnau, who had been the ambassador in London since 1575. He had escorted Walsingham to court to escape the mayhem following the Saint Bartholomew's Day massacre when the Englishman lived in France. Even though Castelnau was a Catholic, he was an old-fashioned ambassador and Renaissance man who considered bonds of friendship more important than the ideological divisions of religion. Even though Walsingham thought of Castelnau as a friend, this did not stop him from trying to recruit a spy to intercept his secret correspondence with Mary.[24]

After Walsingham tried to recruit a former prisoner who provided little information of note, another spy volunteered information about the activities at the French ambassador's house to Walsingham. The spy signed pseudonymously as "Henry Fagot." It turns out one of Castelnau's three secretaries supplied the mysterious Fagot with every letter that Castelnau wrote to Mary for "a little bit of money"; there was nothing he wouldn't let Fagot know, including "all that touches the Scottish Queen and the secret writing in which her letters are written." The treacherous secretary also supplied Walsingham with the names of Mary's chief agents: Mr. Throckmorton and Lord Henry Howard.[25]

This was the straw that really broke the camel's back for Mary's communication channel. Sir Francis Throckmorton, a staunch Catholic, had plotted to murder Queen Elizabeth in 1583 and place Mary on the throne. Walsingham waited about half a year before pouncing on Throckmorton—the conspirator who had been lurking at the ambassador's house late at night. On the night of November 5, Walsingham ordered several of his agents to search Throckmorton's houses in London and Kent and arrest him. When they arrived at Throckmorton's London home, they caught him in the act of enciphering a letter to Mary. Tortured and interrogated on the rack, Throckmorton finally confessed to the plot and to his role in conveying packets of letters to Mary. He was hanged and drawn and quartered in July 1584.

Castelnau had been forced by Walsingham to pass on all the messages to him or be exposed for his involvement in the Throckmorton plot. Castelnau was charged with having general knowledge of the plot and for secretly communicating intelligence with Mary, but the charges against him were dropped. By the fall of 1584, secret communication between Mary and Castelnau had stopped, and a year later he was recalled to France.[26]

Even though the parliamentarians and Walsingham thought that Mary should be executed, Elizabeth overruled them. Walsingham would be able to persuade Elizabeth to execute her cousin only by proving that Mary had been directly involved in such plots. As a result of the Throckmorton plot, security tightened, and on Christmas Eve of 1585 Mary was moved to Chartley, an uncomfortable stone manor house in Staffordshire surrounded by a moat. A new jailer, Sir Amias Paulet, a tough Puritan resistant to Mary's charms, now placed her under strict observation.

Things were looking bad for Mary and her supporters. It was now almost impossible to communicate important information. But this changed suddenly when one of her supporters proposed an ingenious communication system employing beer barrels. Gilbert Gifford, a young Catholic "smooth of tongue as of face, a blue-eyed boy," had approached the new French ambassador Guillaume de L'Aubepine, Baron de Châteauneuf, with the idea. In the sixteenth century, beer, light in alcoholic content, was almost a substitute for water. Chartley, which did not have its own brewing facilities, had to bring in beer for the household. The Catholic brewer agreed to pass on secret letters placed in a waterproof box and stuffed through the bunghole of the beer barrel; the box would then float on the beer. Although the brewer was a supporter of Mary, he had been bribed and continued to betray Mary for Walsingham's money. Both ciphered letters and secret-ink letters written in alum began to be passed back and forth between Mary and her supporters on the outside. And they all landed on the desk of the pockmarked Phelippes for deciphering. Once Walsingham had read the letters, which had been opened by his master seal lifter Anthony Gregory, they were resealed with forged seals, and sent on. But there was a problem.[27]

Gifford had become a double agent. He now worked both for Mary and for Walsingham, whom he had contacted on his arrival in England from France in December 1585. He told Walsingham that he had been acting as a courier between Mary and her supporters on the continent. It is not clear what motivated him to work both sides, but he told Walsingham: "I have heard of the work you do and I want to serve you. I have no scruples and no fear of danger."[28]

Meanwhile, Mary wrote to Castelnau's successor Châteauneuf, warning him about spies who might be in his household "under colour of the Catholic religion wherein your predecessor was much abused." She also urged him to use alum only when necessary, since it was so common and "easy to be suspected and discovered." She advised him not only to write with it in notes carried by her couriers but also to write between the lines of books on every fourth page, or to write on cloth or linen as she had instructed Castelnau. She also advised Châteauneuf to stuff his cipher letters into high slippers instead of cork, because such footwear was common, hence unsuspicious, merchandise.[29]

Even though Mary warned Châteauneuf about spies in his own household, she trusted Gifford, who now became privy to a new plot to invade England, murder Elizabeth, and place Mary on the throne. Anthony Babington, a rich young Catholic who had gone to France and become embroiled in Catholic conspiracies, spearheaded the notorious Babington Plot. After thirteen conspirators had been recruited, Babington proposed six men assassinate Elizabeth. Of course, Walsingham read every detail. This was the moment he had been waiting for. Since no one named the conspirators, Phelippes forged the famous postscript asking for their names. The letter bearing their names was then smuggled to Mary at Chartley via the beer barrel express.

Since the ciphers were simple nomenclators—letters of the alphabet were replaced by numbers, symbols, or Greek letters—Walsingham's code breaker Thomas Phelippes quickly broke the code by using frequency analysis, uncovering the plot and revealing the names of the conspirators. The conspirators, including Babington, were tortured, disemboweled, then hanged. Mary was put on trial and beheaded in the Great Hall in Fotheringhay on February 7, 1587. Her only friend at the end was her small dog, who emerged from her bloodstained petticoat and lay between her shoulders and her decapitated head.[30]

According to Simon Singh, author of *The Code Book,* Mary's "life hung on the strength of a cipher."[31] Mary wasn't the first or the last person who entrusted her life to a weak cipher. But there is more to the story. Imagine what it was like for Mary, isolated in her various moated castles and her manor house. She was to-

The Execution of Mary Queen of Scots, Dutch school, watercolor on paper, ca. 1613

tally and deliberately cut off from the outside world. Her every move was monitored. But Mary desperately needed to communicate with her allies outside of prison, whether by nomenclators or secret ink.

Alum—potassium aluminum sulfate—wasn't a novel invisible-ink substance. It had been used in Italy and described by della Porta. But it wasn't as common as lemon, orange, and onion juice for invisible writing in England. Unlike the acidic juices, alum turned white, not brown, when heated. So even when it was developed, it wasn't as easy to see as the juices.

A finely ground white substance, alum resembles heroin. (I therefore avoid carrying it with me on planes when I give invisible-ink demonstrations.) Readers are probably familiar with alum because it is a pickling spice available at the supermarket. It was an even more common substance in Elizabethan England, because wool traders used it as a mordant, a substance that helps fix dye in cloth. Alum production was already a large industry by the medieval period.[32] Women also used it to get rid of freckles on the skin.

Mary, Queen of Scots, was not the only person in England to use invisible ink to communicate with her confederates. One of Walsingham's most important spies, Thomas Rogers, alias Nicholas Berden, used invisible ink to send secret messages to Walsingham from Paris during the 1580s.

Rogers was one of many scoundrels recruited to spy on the Catholics. He emerged from obscurity as the servant of George Gilbert, a prominent Catholic layman. He first contacted Walsingham in 1583, offering information on Mary. His loyalty to Catholicism had already come to be doubted by other Catholics, who suspected him of treachery. He was then imprisoned at the papal castle of Sant'Angelo.[33] When he was freed, he paid his allegiance to Catholicism but secretly offered his services to Walsingham.

In 1585 Walsingham sent Rogers to France to spy on exiled English Catholics in Rouen and Paris. Rogers sent reports written in invisible ink from August 1585 until January 1586. The visible part of the letter would usually start discussing his search for "parcels" that he was supposed to obtain for Walsingham; the text below, in invisible ink, described Catholic activities. In his first letter of August 11, 1585, he described divisions in the Catholic community. Judging from the brown scorched quality of the paper remaining in the British State Papers, he probably used lemon juice.[34]

After infiltrating these Catholic communities, gaining their confidence, and obtaining their plans, Rogers returned to England and began to compile dossiers on every Catholic priest in the country. He even chillingly recommended hanging or banishment for the imprisoned priests: "Such persons I have noted to be hanged are of most traitorous minds and dispositions. Such as I have marked for banishment are most meet for the said purpose, for that they are exceedingly poor and contentious. Such as I have marked for Wisbech [Castle] are well able to defray their expenses, of the rarer sort and best accounted for learning. And it might stand with the pleasure of his honour, it were meet they should all be hanged."[35]

But there was one way a priest could save his life. Rogers was easily bribed. He tells Phelippes that one priest was worth twenty pounds and the next thirty: "The money will do me great pleasure, being now in extreme need thereof, neither do I know how to shift longer without it."[36]

By the late 1580s Rogers thought the exiled Catholic community might have suspected his acts of betrayal, and he decided to leave the world of shadows to become purveyor of poultry for the royal house.[37]

Arthur Gregory, the master counterfeiter sealer who worked for Walsingham, could also read and use invisible ink. In February 1586, when the beer barrel express was in full force, Gregory apparently had tried to read Mary's secret writing and wrote to Walsingham about his "trials of many ways to discover the secret writing." Despite a swelling in one of his eyes, he found that "the writing with allome is dissolved in divers ways—with fire and with water . . . but most apparently with coal dust thereon, which bringeth it forth white." He also told Walsingham about his experiments with copperas and gall—also a common, and reversible, invisible ink used during the period. Gallnut would reveal copperas, and copperas would reveal gallnut.

But Gregory's most telling experiment was the postscript to his letter, written in alum and instructing Walsingham to "rub this powder within the black lyne the letters will appear white." As the facsimile from the archives shows, Walsingham followed the instruction and the enclosed black coal dust revealed white letters![38]

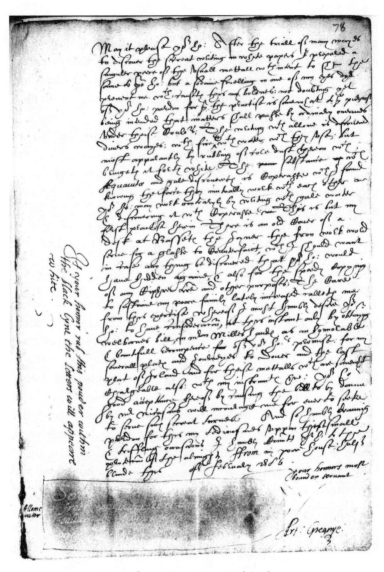

Letter from Gregory to Walsingham

John Gerard's Escape

As the battles continued to rage between Catholics and Protestants in Elizabethan England, the Rome headquarters of the Society of Jesus launched an English mission to support Catholics in England about a year after Mary's beheading and soon after the Spanish Armada was defeated. As part of this mission, John Gerard (1564–1637), a tall, dark, and tenacious English Jesuit priest with a modest and pleasant manner, was sent to England on a covert mission with three other priests in November 1588. His extensive training for the mission at the English College in Rome proved effective up to a point, but then the unimaginable happened.

After Gerard and his team landed on the lonely coast of England, on a moonless, overcast night, they had to sleep in bushes because dogs barked whenever they approached the nearby farmhouse. The next morning, groggy from lack of sleep, the team dispersed and Gerard began a successful six-year undercover mission. His first stop was London, where he picked up instructions from Father Henry Garnet, the superior of the English mission.[39]

Gerard quickly developed a strong Catholic following. It was easy for him to meet new people in the guise of a gentleman of leisure who hunted and gambled and pursued the art of falconry. He charmed his way into Catholic households, which then sheltered him. Many of his new friends never even knew his real identity as a Jesuit priest.

But he also had strong supporters among fellow Jesuits, who hid him at their homes when necessary. The Jesuits even hired carpenters to build hiding places—so-called priest holes—in the rafters, under the floorboards, in fireplaces, or between the walls.

One day a house where he was hiding in Braddocks was searched, but the queen's priest hunters couldn't find Gerard, despite knocking on walls to detect hollowness and ripping out any suspect carpentry. As they searched that house for two days, they sometimes approached Gerard's hiding place. He would hold his breath and they would move on. As his pursuers ripped apart the house, they lit a fire in the fireplace—using logs that had been placed there as a decoy. Gerard had hidden himself in a false bottom under the brick base of the fireplace, which insulated him from the fire. When he emerged, he was faint and tottering because he hadn't eaten for four days.[40]

After narrowly avoiding detection, Gerard moved on to his next hiding place. Unfortunately, this time, the priest catchers caught up with him. He was detained and jailed and shuttled from prison to prison until reaching the dreaded Tower of London. Although his captors threatened torture, he refused to talk. Above all, he refused to name names and locations of other Jesuits, especially his superior, Father Garnet. Soon the lord commissioner threw up his hands with impatience and ordered that Gerard be tortured until he talked.

They took Gerard to a pillar in a huge underground chamber with iron staples driven into the top. After shackling his wrists with iron gauntlets, his jailers ordered him to climb several steps. They lifted his arms, attached a bar to the iron staples, attached the iron gauntlets, and kicked out the wicker steps, throwing the weight of his body onto his arms and shoulders. Although he was paralyzed with pain, he refused to confess. He was tortured three times but still survived, though he couldn't move his fingers for three weeks.[41]

While he was in the Tower of London, Gerard made friends with his prison guard, who began to bring him money and other useful items. Soon Gerard started asking for oranges. He knew that his guard liked oranges, so he gave his new friend some of the fruits as gifts, reserving some for himself. He squeezed the juice into a jar, then made crosses and rosaries out of the peel.

But Gerard didn't drink the juice; he saved it. He was thinking of another way he might use it later.

Gerard knew that requesting a pen would have raised suspicion, so he asked for a quill to clean his teeth. He sharpened the quill and stuck it in a wooden container, and now he was ready to write with his orange juice. He started by wrapping his rosaries in paper, first writing on the wrapping paper. His prison guard gladly sent the material along. In his first message he simply asked his friends to acknowledge that they had read the message. They wrote back on paper used to wrap sweetmeats they sent him. Soon Gerard was regularly exchanging secret communications with his confederates outside prison walls, who later helped him execute a plan for his escape.[42]

As several months passed, the secret communicators grew more confident because the prison guard was very accommodating. (It helped that he was well greased by bribes.) The guard then gave Gerard the pen he asked for. Now Gerard could write secret messages in between the lines of the penned letters.

Gerard preferred using orange juice rather than lemon or citron juice for secret writing because he thought it was more secure. His experience was that lemon juice writing could be revealed with water and heat, but it would disappear with time and could be dipped in water or heated again to reveal the text again. As a result, he wrote, the recipient could not determine whether the letter had been read. By contrast, orange juice writing did not disappear again after it was heated. It could also be revealed only with heat, not with water. The text remained visible and the recipient could determine whether the message had been intercepted and read. (In fact, contrary to Gerard's distinctions, all citrus juices react the same way.)[43]

Meanwhile, Gerard, who never used real names in his letters, started trying to communicate with another Catholic prisoner, John Arden, housed across the garden from his part of the Tower of London. At first, Arden was baffled by Gerard's gestures across the way. Gerard picked up a piece of paper, then pretended to write on it and wave it over an imaginary flame. After persuading his guard to send his fellow inmate a package of rosaries wrapped in paper with a secret message, he waited for a reply. Three days passed; none came. Gerard correctly guessed that the prisoner hadn't understood his gestures.

He tried again. He squeezed an imaginary orange to make juice, and then he dipped the quill in this juice and waved a piece of paper over a flame. Arden caught on. They were in business, and the pair conspired to escape the Tower of London. It was the most daring escape in the Tower's history.

Gerard noticed that Arden's tower had a moat next to it and thought it would be possible to descend on a rope from the roof to the wall next to the moat. He wrote a secret message to John Lillie, his outside confederate, and asked his group to bring a rope to that wall. Arden and Gerard would then throw them an iron ball with a sturdy string attached, to which Lillie could attach a rope. Then Arden and Gerard would pull the rope up to their roof and scale down it. The first attempt was aborted because of unexpected intruders on the riverside, but the second attempt, on October 5, 1597, was successful. Despite his mashed fingers and weakened physical state, Gerard clung to the rope and made it down with his accomplice. A rowboat awaited them, and they rowed off to Father Garnet's house.

There is no doubt that Gerard was an ingenious survivor. Of the four priests who made the landing on the shores of England in 1588, Gerard was the last one alive. Even Father Garnet met his death in the wake of the Gunpowder Plot in 1605. After the failure of that plot, Gerard fled back to the Continent and remained active in the Jesuit community.

Gerard's escape was an amazing success story even though he used relatively simple invisible ink. By contrast, the fate of

Mary, Queen of Scots, hung in the balance between a weak cipher and well-known invisible ink. She was beheaded as a result. A combination of luck, skill, personal charm, and a simple form of invisible ink helped Gerard engineer one of the most successful escapes in history.

Confessing Secrets 3

THE ELIZABETHAN PERIOD, AN age of secrecy and intrigue, had been a tumultuous time for prisoners, scientists, and spies; their livelihoods, even their lives, often hung in the balance between poor ciphers and primitive invisible ink. It is therefore not surprising that there was an explosion of interest in the science of secret communication a century later, during what is often called the Scientific Revolution of the seventeenth century. Secret writing flourished and was used to conceal scientific secrets or secrets of state, or simply to communicate privately. Even though cryptography was more scientifically sophisticated, invisible ink was more magical and mysterious. It was a curiosity, a rarity that raised people's wonder. In the mid-seventeenth century, an anonymous British author vividly described what happens when invisible writing made of vitriol (a sulfate of copper, iron, or zinc) was rubbed with gallnuts: "It will suddenly burst out like lightning from a cloud."[1]

As anyone who has experimented with secret writing knows, invisible writing often *does* burst out like lightning from a cloud. It is magical to watch, a wonder of nature. But during the Renaissance period of intrigue and scientific revolution, it also had a serious purpose. While Jesuits like Gerard, queens like Mary, and spies like Nicolas Berden used simple forms of invisible ink, a century later scientists also began to experiment with more sophisticated types of reactions than fruit juice and heat.

Even though the explosion of interest in secret communication coincided with the Scientific Revolution, little attention has been paid to the connection between the leaders and luminaries of the intellectual and organizational revolution surrounding the birth of modern science, on the one hand, and the science of secret communication, on the other. It may seem paradoxical that leaders of the Scientific Revolution who espoused open communication were intimately involved with the science of secrecy. Many developed new codes, ciphers, and secret inks, and some volunteered their expert knowledge to the state. When we look at the activities and pronouncements of the scientists active during this period, we get a different picture from that of the traditional openness/secrecy dichotomy we are familiar with. It wasn't just an either/or situation for them.

The relationship between secrecy and openness for most of these scientists was a matter less of tension or conflict than of compartmentalization. Despite their advocacy of open and free communication in science, leaders and luminaries of the Scientific Revolution still believed that some things in science and statecraft were better kept secret. And this is why many of them developed novel secret communication methods. They found ways to reconcile secrecy and openness in science and for the state. In fact, the intersection of political intrigue and the birth of modern science can account for, in part, the flourishing of secret communication.

Three major scientists and leaders involved in the British Scientific Revolution were also interested in secret communication: John Wilkins (1614–72), John Wallis (1616–1703), and Robert Boyle (1627–91). All were actively involved in creating the new scientific societies that began to dot the scientific landscape in Europe, and each made important contributions to cryptography or steganography. In addition to their intellectual predilections and talents, their personalities played a role in their attitudes toward secrecy. Wilkins, for example, wrote a delightful and widely read book on the subject.

While the British triumvirate was active in secret communication and the creation of open institutions for science, their European counterparts, such as the Abbott Johannes Trithemius in Germany and the pharmacist chemist Nicolas Lemery in France, were also making pioneering contributions in hidden writing, but they weren't active in founding open institutions for science. A major feature of the scientific movement in Europe was the belief in an ethos of open communication, in contrast to the secret tradition inherent in alchemy, hermeticism, and the craft tradition. Many of the scientists in this movement were also fascinated by the new modes of secret communication, from codes and ciphers to hidden and invisible writing, and even developed innovative methods themselves. By the fourteenth century, alchemists and scientists had adopted secret writing to keep their discoveries secret. But it was not until the Renaissance that interest in the subject exploded, which resulted in an acceleration of advances in cryptography. Now cryptology began to be applied to secrets of state more often than to secrets of nature. Ciphers became almost a new language. Even though invisible ink was less sophisticated, it played a hidden role in secret communication, along with its better-known counterpart cryptography.

The Swift and Secret Communicator

Possessing considerable personal charm and organizational skills, John Wilkins was a reconciler of extremes, whether between science and religion or between secrecy and openness. Wilkins was a founder of the Royal Society, and his interests ranged from theology to science, as did his vocations, from warden of Wadham College, Oxford, to bishop of Chester. In his youth he wrote imaginative books about flying chariots, submarines, and men on the moon, as well as on ciphers and invisible writing. A promoter of the new astronomy, he was a dynamic scientific entrepreneur and popularizer in search of a universal language.[2] In his youth he became fascinated with secret communication methods and published the first book on ciphers and invisible writing in English, *Mercury; or, The Secret and Swift Messenger* (1641), at the age of twenty-seven.

Unlike della Porta, who loved a secret, Wilkins sought openness and was a strong believer in public knowledge. As a founder and first secretary of the Royal Society, he espoused open communication among scientists and the dissemination of knowledge for the good of mankind. He always wrote his books in English instead of Latin, in order to reach a wider audience. His popular science books were the most widely read scientific books in his day.

Born in the small town of Fawsley, Northamptonshire, Wilkins was the son of a goldsmith and grandson of a vicar. After studying the classics, mathematics, and astronomy at Magdalen Hall, Oxford, he served as a vicar in his hometown before his appointment as warden of Wadham College, Oxford, in 1648 by the Puritan political activists known as the Parliamentarians. Wilkins was a bon vivant who organized lavish "entertainments"; John Aubrey described him as a "lustie, strong growne, well sett, broad shouldered" man. He was, in fact, slightly corpulent, with curly blond shoulder-length hair, sly eyes, and a friendly demeanor.

Long a bachelor devoted to his circle of scientific friends, Wilkins, at age forty-two, married Oliver Cromwell's sister Robina French, twenty years his senior. Bishop Burnet defended

Wilkins's marriage arrangement: "He made no other use of that alliance but to do good offices." Cromwell personally lifted the ban on marriage in the priesthood for this case in 1656. Whitehall Palace, the official residence of the lord protector since 1654 and home to government offices, became the Wilkinses' "marital accommodations." Because of this new alliance he was named master of Trinity College, Cambridge, in 1659 by Richard Cromwell, but quickly lost the position the next year with the Restoration.[3]

Wilkins and his scientific friends liked to meet at the Bullhead Tavern in Cheapside to discuss experimental philosophy and have a pint; when the group became too big for a club, they moved into a parlor at Gresham College. Wilkins was "heartily loved by all" and considered a good man who supported other people. Bishop Gilbert Burnet spoke with great warmth about Wilkins when he said he was "as good a soul, as any I have known. . . . He was a lover of mankind, and had a delight in doing good."[4]

But long before he started hobnobbing with the Parliamentarians, young Wilkins was inspired to "collect . . . notes" on secret communication after reading a pamphlet, *Nuncius Inanimatus*, by Bishop Godwin on communicating at a distance without a messenger through fire and signs, whether the friend was in a "close dungeon, in a besieged City, or a Hundred Miles off." These feats "raise[d Wilkins's] Wonder."[5]

After he had collected enough material to satisfy his curiosity, Wilkins wrote *Mercury; or, The Secret and Swift Messenger,* to his "own farther Delight." It was the first major book in English on secret communication and remains one of the best overviews of ancient methods. In the first edition he admits that he published the book to gratify his brother, a printer, and that "the vanity of this age is more taken with matters of curiosity, then those of solid benefit." Published a year before the Civil War started in 1642, *Mercury* was timely and useful to all sides of the political or religious divide.

Wilkins found some ancient methods strange and primitive. He assures the reader that during ancient times the methods must have appeared more secret than they do to the modern eye. He writes that nothing can be compared to the "strangeness" of shaving a slave's head and writing a secret message on it, waiting for the hair to grow, and sending the slave to his destination, as described by Herodotus. Clearly, this method was horribly slow. For Wilkins, modern methods of conveying messages— using inanimate media like bullets and arrows or animate media like men, animals, or birds, or sound, or even angels—were swifter. His attitude toward the ancients was reminiscent of the quarrel between the ancients and moderns: "You may see what strange shifts the Ancients were put unto, for want of Skill in this subject," he wrote.[6]

Wilkins coined some new cryptographic words and included several lesser-known codes and ciphers, but his treatment of invisible ink relied heavily on della Porta. He does tell us there are "sundry Ways of Secrecy" in the material of writing—the ink or "Liquor" used. We learn again about "Salt Armoniack" and lemon and onion juice. Wilkins was interested in how secret messages were transmitted on land, through the water, or in the air and was very impressed with della Porta's incredible ability to discover secret writing.

Wilkins had no qualms about revealing secrets on how to conceal secrets; he was aware that his book taught not only how to "deceive," but also how to "discover Delusions." Though the god Mercury had the reputation of being a thief and traitor, Wilkins squelched fears that his eponymous book could be used for unlawful purposes. His attitude was that "not . . . every thing must be supprest which may bee abused. . . . If all those usefull inventions that are liable to abuse, should therefore be concealed, there is not an Art or Science, which be lawfully protect." He added graphically: "Wee may as well cut out our tongues because that member is a *world of wickedness.*"[7]

Wilkins and the Royal Society

One of the best-known societies for the open communication of science was the Royal Society, and John Wilkins was instrumental in its creation. Before the charter founding the Society in 1662, science at Oxford had thrived under Wilkins's wardenship at Wadham College, 1648–59. Wadham became the center for scientific studies in England during the 1650s, and Wilkins attracted the best scientific minds to Oxford. Wilkins's Oxford Club is often considered the Royal Society in embryo.[8] Many of the members of his group, and fellows of the Royal Society afterward, like Robert Boyle, Robert Hooke, Christopher Wren, and John Wallis, had an interest in secret writing—a practical interest in some cases. During the Oxford years, amid the myriad of experiments, the group developed "new ways of Intelligence, New Cyphers," for themselves and for the state.[9]

Although open communication of science was part of the Baconian ethos institutionalized at the Royal Society, some things were still kept secret there. In fact, the origins of a notion of national security seem to have begun in seventeenth-century England. The president of the Society either was entrusted with secrets or made an executive decision to keep a secret. When William Petty presented a manuscript on shipbuilding to the Royal Society in 1665, Lord William Brouncker, the president, took it away and kept it because he said it was "too great an Arcanum of State to be commonly perused." That same year, Robert Hooke announced at a meeting that he planned to place "his secret concerning the longitude into the hands of the president, to be disposed of as his lordship should think fit."[10]

Scientists at the Royal Society also used ciphers or anagrams to keep their secrets for future use or for priority reasons. In 1669, Christian Huygens sent Secretary Henry Oldenburg a "cipher . . . to be lodged" in the Society's register book. He also proposed using anagrams as a "way of securing his discoveries or inventions." Five years later Huygens sent Oldenburg "a new invention of watches" concealed in an anagram. A contentious priority dispute arose between Huygens and Hooke when Huygens claimed that he had developed a watch for longitude long before Hooke.[11]

Some Society fellows became gun-shy about presenting results at the Society because the information leaked to the society at large, and others could steal their ideas. To rectify this, the Society granted "secrecy to members of the Society," and all observations, experiments, inventions, and discoveries presented there could be "kept secret at the desire of the communicator." The knowledge was not allowed to leave the meeting rooms.[12]

Henry (Heinrich) Oldenburg (1619–77), the Society's secretary responsible for foreign correspondence, was discrete about secrets. Known as the philosophical "intelligencer," he maintained an enormous network of scientific correspondents, spanning the Continent, and he hauled in many scientific secrets for his British colleagues at the Royal Society.

Born in Bremen, Germany, Oldenburg was educated as a theologian but came to England as a diplomat and stayed on as a seemingly loyal subject. An original fellow of the Royal Society, he founded *Philosophical Transactions* and originated peer review of manuscripts.[13]

Given his murky background and his network of correspondents, it is not surprising that he was arrested and imprisoned in the Tower of London in 1667 during the second Anglo-Dutch War under suspicion of espionage ("for dangerous designes & practises"). He was detained for only a month.[14] Modern intelligence agencies would love to recruit an active networker like Oldenburg!

But his story has a strange twist. Joseph Williamson, the undersecretary of state who threw Oldenburg into prison, used him as an agent to obtain political information. Oldenburg was even willing to have his mail diverted to Williamson's office for examination before it was delivered.[15]

The Invisible Man

While Wilkins was a bon vivant who promoted the new "experimental philosophy" and organized the Royal Society's scientists, Robert Boyle was an ascetic devoted to experimental philosophy, known for his work on the air pump and for formulating Boyle's law (the observation that pressure and volume of a gas are inversely proportional). His politics were not as overt as Wilkins's or Wallis's; in fact, it is not obvious what side of the political fence he stood on. What is clear is that he followed the advice of his father to be cautious; he remained neutral, siding with neither the Royalists nor the Parliamentarians. If anything, he embraced the moderate Latitudinarian point of view.[16]

He was tall, about six feet, slender, and pale, and he possessed a long face and flowing, curly hair. Aubrey describes him as "very temperate, and vertuouse, and frugall: a batcheler; keepes a Coach, sojournes with his sister, the lady Ranulagh. His greatest delight is chymistrey." His temperament and strengths complemented Wilkins's.[17]

The fourteenth child of Richard Boyle, the richest man in Great Britain, young Boyle was sent to the Continent for a European tour with a tutor after he completed Eton College at the age of thirteen. His interest in the sciences seems to have been sparked by his time in Italy, where he learned about Galileo's discoveries. Upon his return to England at the age of nineteen, he lived with his sister Lady Ranelagh, who introduced him to Wilkins and his circle in Oxford. Wilkins was very impressed with Boyle's promise and warmly welcomed him into the group and encouraged his budding scientific efforts. The lonely young Robert Boyle was exuberant about the philosophical group that "honored" him with its members' company, and affectionately dubbed them the "Invisible College" in 1646.[18]

Like his contemporaries Robert Hooke, Galileo Galilei, and Christian Huygens, who concealed their scientific writings through anagrams or simple ciphers, Boyle kept hundreds of pages of his private notes and letters obscure through word, name, letter, and numerical substitutions. Lawrence M. Principe has argued that Boyle kept his alchemical notes secret using ciphers and codes (no mention of invisible ink) because of their special secretist status, and to prevent assistants and amanuenses from reading or stealing.[19] But Boyle had other reasons for keeping secrets as well: their use for barter, because of trade secrecy and financial reasons and for secrecy of state or mankind. He tells us: "I have not yet thought fit so plainly to reveal, not out of an envious design of having them bury'd with me, but that I may be always provided with some Rarity to barter with those Secretists that will not part with one Secret but in Exchange for another."[20] He received some trade items "upon Condition of secrecy," or because when they were communicated to him or imparted by him, he "[did] yet make, and need to make, a pecuniary advantage of them." Boyle spared no cost "to get any rare secret."[21]

Boyle has typically not been included in the pantheon of Renaissance secret communicators, yet I have found that he is the first scientist to actually use the English-language term "invisible ink" in print, in 1665 (and in manuscript form many years before). Boyle "concealed" most of his work on invisible ink (and poisons) "for the Good of mankind" and because he thought it was "mischievous."[22]

Boyle experimented with both organic substances and more sophisticated chemical combinations. His organic methods included using urine and blood as invisible ink. After he observed that human urine "would tolerably serve for what is call'd an invisible ink," he thought he might try the serum of blood because of the "suppos'd . . . great affinity" between the two substances. In the experiment he described, he and his colleagues took some serum of blood, "dipt a new pen in it, [and] trac'd some characters upon a piece of white paper." When it was dry they held the unwritten side of the paper over a flame of a candle. The letters appeared. Though they were not inky black, they were dark enough to be legible. He found the results similar to those with urine.[23]

Boyle also developed more sophisticated methods, including one in which the invisible message was written on top of the visible cover letters. The "liquor" would erase the cover writing and develop the invisible writing. He also knew that "appropriate liquors" were needed to make certain invisible inks "confess their secrets," a wonderful turn of phrase describing the action of a reagent.

This remarkable invisible ink actually used arsenic sulfide, known in the seventeenth century as orpiment or auripigment (the latter for its gold color), as the main ingredient "to confess its secrets." Magicians used it later in a modified form to enthrall their audiences with the magic of a vapor penetrating the pages of a whole book to reveal invisible writing on the other end.

Boyle experimented with different combinations incorporating arsenic. In 1665 he tells his reader to start with three parts of quicklime (calcium oxide, a caustic alkaline) and one part orpiment (yellow arsenic sulfide), "beat the lime grosly," and "powder the orpiment (with care to avoid the noxious Dust that may fly up.)" Next, Boyle writes, these ingredients should sit in water for two or three hours, with occasional agitation. While this mixture settles, he writes, the reader should make a black ink of burned cork mixed with gum Arabic. The next step is to create the invisible ink with powdered red-lead (lead acetate will work) and vinegar. Those substances should infuse over hot coals for a couple of hours until the "liquor" acquires a "sweet taste."

Boyle liked to write the invisible message first with a "clean Pen," then write over that, when it was dry, with regular black ink. To develop the invisible ink Boyle dipped a sponge or "Linnen-rag" in the filtered solution made of quicklime and orpiment. Though "it smell ill"—arsenic sulfate can smell like rotten eggs or sewer vapors—when the clear "liquor" is wiped over the black and invisible ink, it obliterates the black and develops the invisible "at once."[24]

But the real magic came later when Boyle mixed a solution of sulphur, quicklime, and sal ammoniac (ammonium chloride)—they have the same effect as the first developer—in a beaker. He lay a piece of paper with an invisible-ink message on top of the vial, and the vapors from the solution developed the message. It also developed a sheet of paper hidden within a stack of six pieces of paper when the exterior of the bundle was exposed to the vapors. Boyle explained this by his notion of effluvia— invisible emanations, usually noxious smelling—and adamantly rejected any notion of occult processes.

In France, Nicolas Lemery (1645–1715), a chemist and pharmacist, became interested in "inks called Sympathetical," especially those employed in the experiment described by Boyle. Lemery's claim to fame was the popular best-seller *Course of Chemistry* (1675), which went through thirteen editions in his lifetime and became a standard work translated into several languages. It sold like a book of romance or satire and catered to Parisians' interest in popular science. Even though the text focused on practical chemistry for the doctor, in which context Lemery

considered the arsenic recipe "of no use," he called it "surpriz-
ing" and called for an explanation. Because the lead acetate and
arsenic-quicklime recipe seemed to work by "sympathy"—an
unknown cause—invisible ink was soon called "sympathetic
ink" (*encre sympathique*) in France, terminology that spread
rapidly to the rest of Europe. But Lemery was eager to look for
an explanation "without having recourse to *Sympathy* and *An-
tipathy*," which he thought were "general terms" that explicated
"nothing at all."[25]

Lemery was one of the first scientists to develop theories
about acid-base interactions, and these theories helped him explain why the black visible ink became invisible with the application of the quicklime and arsenic and why the invisible ink
became visible with the same solution. The key to Lemery's explanation is that the black sooty substance on the burnt cork
used to write the message is oily. When the quicklime alkali-
and-arsenic mixture is applied, it acts like a soap and washes the
sooty substance away. Conversely, when the quicklime-arsenic
reagent is applied to the invisible lead, which has suspended acid
on its edges, it is revived by the mixture of alkali quicklime and
sulfur of arsenic and neutralizes the acids. Lemery explains that
the reagent develops the invisible ink on the other end of the
book because the sulfur penetrates the pages.[26]

Boyle wasn't the first person to experiment with these magical inks. In fact, the arsenic experiment may have originated
with French iatrochemists, who were practical-minded physicians or pharmacists who applied the useful results of chemistry to medicine or pharmacy. One of these iatrochemists took
this experiment one step farther. He referred to his discovery as
"waters that act at a distance."

Pierre Borel (1620–78) heard about this secret from a Montpellier doctor and, like Boyle, traded the rarity for another one.
Borel was a French physician, naturalist, and chemist who stud-

ied in Montpellier, where he became a doctor. After moving to Paris, he became physician to King Louis XIV and a member of the Royal Academy of Sciences. He collected enough curiosities to fill his own private museum and published a catalogue of these rarities. It is no wonder that he was fascinated by "magnetic water that acts at a distance." In 1653 he described the way in which metallic solutions became visible when they were exposed to the "action of certain vapors."[27]

Even though it is not clear whether Boyle had heard about Borel's experiment, he had heard about a similar experiment by another Frenchman who lived in Italy, Claude Berigard. In fact, in 1657, Henry Oldenburg, secretary to the Royal Society, sent Boyle a recipe for "a way of writing to others very secretly" because he thought Boyle did not "have . . . this way." Oldenburg relates that an Italian who was teaching Boyle's nephew practical geometry and fortification refused to exchange the secret for another one. But money ("more powerful than the thunderbolt, it breaks down every barrier") changed his mind. Oldenburg thought the "receipt" would be of "great use" for "besieged towns"; he even tried it out himself and it worked.

The recipe was written in French and was similar to the one Boyle described in 1665: It used a mixture of vinegar and litharge (a lead oxide used in paints and enamels) as the invisible substance. The recipe called for burning a piece of cork until it was carbonized, pouring brandy over it, and mixing gum to make an ink to write over the white vinegar–litharge blend. Then one mixed one ounce each of orpiment, quicklime, and water. Presto! This solution would "efface" the white letters and make the message visible just as in the experiment Boyle had described.[28]

The arsenic invisible ink recipe continued to be reproduced well into the late nineteenth century in chemical recreation and recipe books. People were enthralled by the vaporous penetration of the orpiment solution to the other end of the book. Invisible ink was already something magical, but penetrating the pages of a book or even a wall was supernatural! Recipes instructed the experimenter to insert one page between the first pages of a volume and another sheet brushed with the orpiment mixture at the back of the volume. This is the way a writer from 1753 dramatized it: "Shut the Book nimbly, and with your Hand strike on it two or three smart Blows . . . or sit upon it for a few Minutes; after which, on opening the Book, you'll find the invisible Writing, black and legible, by the subtile Penetration of the Steams of the Orpiment through all the Leaves." Like others, the author warned that the experiment "should be made in the open Air and with great Caution, the Fumes of the Orpiment stinking most abominably, and being productive of great Mischiefs if taken into the Lungs."[29] It is no wonder that Boyle was hesitant to share such secrets with the government.

The Cipher Breaker

In contrast to Wilkins and Boyle, John Wallis, Savilian Professor of Geometry at Oxford, was publicly willing to serve his country by breaking ciphers. A man of medium build—heavyset in later years—he had a somewhat round, small head and short, simple hair. He began his career as government code breaker quite by chance. After receiving his bachelor's (1637) and master's (1640) degrees from Emmanuel College at Cambridge, where he studied theology and mathematics, he became, at the age of twenty-six, chaplain to the widowed Lady Vere at the beginning of the Civil War in 1642.

During dinner at Lady Vere's house in London, a chaplain to Sir William Waller, a member of the Long Parliament, showed Wallis a letter in cipher related to the capture of Chichester. The dinner guest showed it to him "as a Curiosity . . . and asked

. . . between jeast and earnest" whether Wallis could decipher it. Wallis accepted the challenge and deciphered the letter after dinner in about two hours. It was the first of many letters he deciphered during the Civil War and afterward. It was also the first time he ever even saw a cipher; he had not even read any of the literature on ciphers. When he finally looked at della Porta's book, he found it of no use because it described only how to write a cipher, not how to break one. By the twilight years of his life, however, Wallis's successes declined as work by Blaise de Vigenère made French cipher methods more complicated. Wallis was rewarded for his early successes with his appointment to the chair in 1649 (replacing a Royalist), a few weeks after England's transformation into a republic.[30]

Loyal to his country, Wallis fended off Baron Gottfried von Leibnitz's persistent demands that he teach a group of students from Hanover cryptographic methods lest they die with him. Instead, Wallis revealed his methods only to his son and grandson, William Blenow, who later became official decipherer for the Royalists after Wallis's death.[31]

Though loyal to his country, Wallis was apparently willing to serve any regime. Even contemporaries criticized him for deciphering the captured Charles I's correspondence after the Battle of Naseby during the Civil War, thus leading Charles to his death. Wallis denied these accusations.[32] Contemporary historians would probably label such behavior as opportunism, though Wallis may simply have been politically naïve and found the cipher breaking "technically sweet."

Though he complained about not receiving enough compensation for his efforts, his honorarium increased substantially over the years. If the monetary award was not enough, he was also rewarded with the appointment of chaplain to the king during the Restoration. In order to increase his earnings, he reminded the government of the political value of his services when he caught Louis XIV asking the king of Poland to declare war on Prussia: "The deciphering of some of those letters [has] quite broken all the French King's measures in Poland."[33]

Like Boyle, Wallis was part of the Invisible College and Wilkins's Oxford group and also took part in founding the Royal Society. Wallis is known as one of Great Britain's greatest mathematicians, who developed the symbol for infinity (∞), as well as providing Isaac Newton a springboard for his work on the calculus through the *Arithmetica Infinitorum*,[34] yet little has been written about his consultant deciphering work for the British government. Part of the reason for this neglect is the embryonic nature of intelligence organizations at the time and the lack of formal cryptographic branches, as well as the late blooming of the field of intelligence history. Despite this neglect, by the early eighteenth century there was a proliferation of books on deciphering. Although John Wallis did not live to see them, he left his legacy with his son and grandson, who unfortunately committed suicide in 1713.

John Wallis's career exemplifies most clearly and directly the way in which political intrigue and scientific advance intersected in seventeenth-century England. He was also aware that there was an explosion of interest in ciphers. In earlier periods, he writes, ciphers were "scarce known to any but the Secretaries of Princes . . . but of late Years, during our Commotions and civill Wars in England, grown very common and familiar."[35] Even though invisible writing also became "common and familiar," it didn't develop as much as ciphers, in part because mathematics was more sophisticated at the time and opportunistic scientists like Wallis were more willing to work for the state.

Even though invisible ink wasn't as highly developed, Robert Boyle's description of the process of invisible messages "confessing their secrets" through a reagent evokes images of confession in a religious sense: an admission of sin. Secrets, then, connote wrongdoing, and confession brings relief. It is a nice metaphor also for the tension between secrecy and openness.

Secret communication is at the heart of keeping secrets. From symbols and signs to scrambled words and invisibility, the materiality of the medium itself is the secret. In steganography, the ink or the paper is the secret; it began as a material science. Once the "liquor" "confesses its secrets," the hidden is exposed. The invisible becomes visible, the obscure is deciphered. The unknown becomes known. Betrayal is unmasked.

German Secrets

While many British scientists who were interested in secret communication were vocal about the relative merits of secrecy and openness, their French, Dutch, and German counterparts were not science organizers like the British and seemed disinclined to discuss the matter beyond their service to their royal patrons, who were fascinated by the subject. In fact, many on the Continent still lived in an age of secrecy comparable to the Elizabethan Era. But this does not mean that they did not make major contributions to the study or use of secret communication.

Long before the British scientists had become interested in secret communication, Johannes Trithemius (1462–1516), the German abbot, cryptographer, historian, and occultist, coined the term for hidden writing, entitling his book *Steganographia,* which was written for royal patrons, the "princes." Although Trithemius completed the work in 1499, it was not published until 1606 because he was accused of dabbling in black magic. But it circulated widely in manuscript form among other occultists and people interested in secret communication. Unlike the *Polygraphia* (1518), a work on cryptography, which made important contributions to polyalphabeticity, the *Steganographia* was denounced as a work of magic and sorcery and placed on the index of forbidden books in 1609. Ever since then Trithemius has left us guessing: is the bewildering *Steganographia* about cryptographic techniques disguised as angel magic, or is it a magic book disguised as cryptography?

Not until 1998 did an AT&T lab scientist, Jim Reeds, decipher the mysterious book 3 and make the bewildering texts more understandable. By cracking the code in the texts, Reeds makes Trithemius's work look more like cryptography than demonology.[36]

Ostensibly Trithemius explains how to send long-distance secret messages through angels. We learn the angels' names and ranks, and the compass directions, times of day, planets, and constellations associated with them. A spirit's help is apparently obtained by texts of invocations that make up most of the book. But Trithemius was really hiding secret messages in these bizarre texts. He was more interested in concealed cryptography than in what we now refer to as steganography.

He sometimes concealed his messages in the long invocation of angels' names. In one case, he spells out the message by using every other word to highlight every other letter. So

padiel aporsy mesarpon omeuas peludyn malpreaxo

reveals "prymus apex," which could mean "the first to reach the top." But it's not clear what the secret implication is here.[37]

If the *Steganographia* is disappointing because it isn't about contemporaneous knowledge of hidden and invisible writing, the *Polygraphia* offers some new twists on invisible writing. Drawing on his love of language, Trithemius coins the terms "philophotos" and "misophotos" to mean writing that "likes the light" (photos), which we can see, and writing that is hidden (doesn't like light).[38]

Trithemius's influence was long-lasting, and he is often credited as the founding father of hidden writing because of the title of his book. In fact, around the same time, interest grew in steganography, however named and practiced. Sometimes it was the study of codes and ciphers and other times the study of hidden, invisible secret writing. A large group of German Jesuits, in particular, practiced and wrote about steganography.

Athanasius Kircher, the prominent seventeenth-century German Jesuit who wrote about secret writing, has been called the last man who knew everything. Kircher was an occultist whose knowledge ranged from geology to mathematics to medicine to oriental studies. Because of the range of his interests he can be considered the Leonardo da Vinci of Germany. Kircher came of age during the Thirty Years' War (1618–48) and had to flee Germany because of Protestant persecution. Scholar Nick Wilding believes Kircher's interest in creating a universal language was influenced by the fragmented society in which he lived and a "gesture toward reunifying the Holy Roman Empire" in the wake of the war.[39]

Kircher was a firm believer in the importance of secrecy in affairs of state. Since knowledge is power, he thought it essential that rulers conceal or control knowledge. Like della Porta, he believed that princes should keep certain types of knowledge and advice secret. But as with his British counterparts there was a tension between openness and secrecy in his work. This tension, even self-contradiction, is very well captured in the caption to the image of the Egyptian God Harpocrates printed in one of his books on Egyptology. While the God displays the sign of secrecy by placing his finger over his lip, the caption says: "With this one I disclose secrets."[40]

Kircher was also interested in all sorts of different kinds of steganography—magnetic, musical, graphic, and numerical. Some of his work on this subject was similar to Trithemius's in the sense that he believed in communicating at a distance, but his interest also lay squarely in contemporary fascination with magnetism. He developed a machine that functioned like a magnetic telegraph. The device consisted of a row of spherical bottles with magnetized stoppers decorated with dancing angels. Reminiscent of the arsenic penetrating invisible ink, when the bottle with a letter is turned to the next one, the magnetic stopper touches it and passes on the letter by "sympathy" until it gets to the end. Like many other scientists during this period, Kircher was sometimes a natural magician steeped in the world of occult qualities, and sometimes a modern scientist who portrayed a world run by reason.[41]

Other Germans living during the Baroque period were also enchanted by the world of secret writing and messaging. Johannes Balthasar Friderici's beautiful *Cryptographia; or, Secret Writing* appeared in 1684. Although it is considered a classic in its field and includes information on ciphers, gestures, signs, music, and most notably invisible ink, little is known about Friderici except that he was probably from Leipzig and worked as a lawyer in Hamburg. The name itself might be a pseudonym. It is clear, though, that he was fascinated by secret writing and developed or commissioned a beautiful frontispiece with the God of Messaging Mars flying through scenes of codes and invisible writing. But he cited no sources, inviting accusations of plagiarism as early as 1689. Even so, the overview of invisible ink is the first of its kind in the German language; it is more systematic and organized than della Porta's and adds some new material.[42]

Of Steganography, Sympathetic Ink, and Invisible Ink

Although Trithemius gave the modern field of steganography its name, during his time and up through the nineteenth century the term was never used to describe invisible ink or hidden writing as practiced by the ancient Greeks. It was often used interchangeably with cryptography to describe the study of codes and ciphers; it was sometimes also defined generally as "secret

Frontispiece, Friderici's *Cryptographia*

writing." An attempt to read Trithemius's *Steganographia* calls to mind a definition of steganography from 1753 (repeated into the nineteenth century): "a mysterious and unintelligible way of writing." By the 1960s the term was more widely used to refer to all sorts of hidden writing, from invisible ink to microdots, and by the late twentieth century it also encompassed digital hidden writing and watermarking. Friderici was one of the first Germans who started the trend of referring to invisible ink as "hidden writing."[43]

From ancient times to the early part of the Renaissance, most people did not use any term at all to describe writing invisibly. They often referred to writing revealed when held over heat. In a postscript to a 1562 letter from Rouen to England, a British author writes: "Warm this well to the fire and you shall see more.... Warwick should leave no piece of paper untried by fire that the writer sends."[44]

Mary, Queen of Scots, told her confederates to "write in

white" and sometimes called the technique "secret writing," as did Arthur Gregory, Walsingham's master counterfeit sealer. Forever the trailblazer, della Porta was one of the first to use the Latin equivalent to "invisible writing." By the time of the Scientific Revolution, scientists like John Wilkins would refer to secret writing through "waters," and the "ink" or liquid was called "liquor."

It wasn't until the late seventeenth century that phrases like "invisible ink" and "sympathetic ink" began to be used more widely. The new terms reflected how scientists began to explain the wondrous phenomenon. When Robert Boyle wrote of "what they call an invisible ink," he apparently was borrowing from another source. He was also heavily steeped in an invisible world surrounded by his effluvia and theology. By 1693 a British diplomat told a correspondent he could use "the invisible Ink which Mr Beintema gave M Heemsherke. He will give me the Secrett to understand you."[45]

While the British seemed to favor the newly minted term "invisible ink" during the Scientific Revolution, the French began using the phrase "sympathetic ink" (*encre sympathique*). The notion of sympathy originated as a medical term: supposedly, one part of the body induced an effect in another part, usually because of a disease. As chemists tried to understand what was happening when an invisible fluid became visible by heat or another chemical, they sought explanations. By using the word "sympathy," they were expressing the idea that one action could affect something else. But explanations that used the notion of sympathy and occult qualities were often invoked when people didn't have a cogent explanation. In the late seventeenth century and early eighteenth century, sympathetic inks were described as "such as can be made to appear or disappear, by the Application of something that seems to work by Sympathy."[46] This is around the time the new term was coined first by the French. It was tailor made for magicians.

Invisible Landscapes

IN THE CHARMING MOUNTAINTOP village of Schneeberg, Germany, shop windows are filled with hand-carved nutcrackers, smoking men, candelabras, angels, and pyramids. Synonymous with Christmas, this alluring folk art also depicts a rich mountain tradition. The miner's candelabra portrays typical Schneeberg occupations: lace makers who sewed the window curtains that adorn the town's baroque buildings; two miners in knee-high britches, knee pads, and tall hats, carrying a hammer and pick; an angel presiding over the scene; and a wood-carver on the right. The lit candles on the arch represent the miner's most precious commodity: light. Even as late as 2010, when I visited Schneeberg and retraced the miners' steps, villagers still used the miners' greeting *Glück auf!*— godspeed—and the hammer and pick symbol still decorated the town.

It is easy to imagine the music of Bach wafting through the cobblestone streets of this Baroque village. Located in the Erzgebirge, the Saxon Ore mountain range, the Schneeberg Mountains—really just hills of 1,450 feet elevation—contained rich sources of silver, bismuth, and cobalt hidden beneath the bucolic exterior of verdant hills and valleys. The discovery of silver in the mountain in the fifteenth century led to extensive mining of the precious metal and to the founding of a miner's

Diorama of underground mining

village in 1470. Once they depleted the known silver veins, villagers began to mine cobalt more than a century later and removed hundreds of thousands of kilos from the mountain once they recognized its importance. Streets like Silver Road and Cobalt Road snake through the Schneeberg landscape and define its past. While silver brought riches to the town, cobalt produced many unintended marvels, like the magic of invisible ink.[1]

But extracting cobalt was not an easy business during the medieval period (preceding the mining spree), when spirits seemed to inhabit forests, caves, and mountains. The German word *Kobold* means goblin, and a miner's kobold was a mischievous spirit that haunted subterranean places like mines. They were often blamed for causing falling rocks and explosions and even bewitching miners. Kobolds are invisible but are often portrayed as small gnomelike figures. As the yield from the silver

mines in Saxony declined, kobolds were blamed for stealing silver and leaving behind useless rock. Cobalt looked like silver
ore, but it could not be smelted easily and was seen as worth-
less and poisonous because it was laced with arsenic and sulfur.
The story goes that a young smelter arrived in Schneeberg and
started experimenting furtively at night with the dead rock the
miners thought was silver ore. Just as villagers were about to arrest him for wizardry, he discovered how to smelt the material
into a brilliant blue color we know as cobalt blue.[2]

Although the brilliant cobalt blue color had been used since
the time of the Chinese and Egyptians to paint porcelain and
vases, the first large-scale mining and smelting operation began
in Schneeberg around 1600. Soon after, the Blue Color Works
were established to transform the brute rock into the beautiful translucent blue cobalt, dubbed Saxon blue, a highly prized
color often denoting royalty.[3]

Theophrastus Phillippus Aureolus Bombastus von Hohen-
heim (1493–1541), known as Paracelsus, was the first to describe
cobalt from the mines at the border of Saxony and Bohemia—
the Ore Mountains. A physician and renaissance man who was
described as bombastic and also dabbled in alchemy, Paracelsus
occupied a world filled with demons, nymphs, elves, gnomes,
dwarves, fairies, and kobolds. He was not alone. During the
sixteenth century the belief was widespread that nature was
filled with spirits that were associated with mountains, caves, or
mines.[4] Most alchemists at the time believed that spirits inhabited nature. They were also assiduously looking for the Philosopher's Stone, a substance that had the power to turn base metals,
especially lead, into gold.

In 1705 a mysterious female alchemist was reportedly the
first to identify bismuth-cobalt as a substance from which to
make sympathetic ink. She was also the anonymous author of
three alchemical books, including one with the alluring title *On*

the Key to the Cabinet of the Secret Treasure Room of Nature,
which included a discussion of the changing bismuth-cobalt colors. Ever since the book was first published in 1705, people have
speculated about the identity of the mysterious DJW, from Weimar, named on the title page. A few people thought DJW stood
for a "Doctor Jacob Waitz" from Gotha, but most pointed to an
alchemist probably named Dorothea Juliana Walchin, who was
equally mysterious.[5]

Dorothea Juliana Walchin (perhaps Wallich or Wallichin) is
variously portrayed as a daughter of an adept (an alchemist), as
the widow of Johann Heinrich, or as a Saxon woman experienced in chemistry. Hermann Fictuld (a pseudonym for a Baron
Johann Friedrich von Merstorff), a Rosicrucian and Mason, and
head of the Golden and Rosy Cross group, criticized her works
with "great severity," describing them as "ertz-sophistical and
arg-chemical" (terrible), fit only for burning; he warned people
not to read them.

The mysterious woman alchemist seems to have discovered a
cobalt mineral, which was red in a solution and was supposed to
be the "first matter," the Holy Grail of the alchemists. The cobalt mineral also displayed magical qualities: the color changed
from rosy red to grassy green to sky blue when heat was applied. When it cooled off, the color faded back to red. When the
cobalt was prepared and turned into a solution to write with, it
was clear, but it produced that fabulous blue-green color when
heated. The writing disappeared when cooled.[6]

Georg Ernst Stahl (1659–1734), a well-known German chemist and physician, knew Wallich and admired her intellect. She
had been the godmother to one of his children when he lived
in Weimar. He thought she had more experience in chemical,
or alchemical, matters than a great number of learned men. He
knew that she had gone to Schneeberg and experimented with
the cobalt people there. During a time when women were often

not recognized as worthy of praise in the sciences, he came to her defense.[7]

Jean Hellot (1685–1766), the scientist who did more than anyone else to investigate, promote, and advance the subject of sympathetic ink during the eighteenth century, also referred to a mysterious person when allocating credit for the discovery of the magical qualities of cobalt. Hellot acknowledged an unnamed German artist from Stolberg (near Schneeberg) who showed off the wonder of cobalt at a demonstration at the Academy of Sciences in Paris. Hellot lived most of his professional life as an industrial chemist during the early period of the French Enlightenment. A short and rather chubby man, Hellot had eyes that sparkled with liveliness. In spite of this vivacity, he was not soft when it came to commerce or maintaining the interior of his house. Even though some of his written work was rather dry, he was extremely pleasant in conversation.[8]

Though not as famous as his countryman Antoine-Laurent Lavoisier, who revolutionized chemistry later in the century by proving that oxygen, not phlogiston, supported the combustion of flammable materials, Hellot was a pioneer in the technical chemistry of dyeing, mining, and assaying. His best-known work, on the art of dyeing wool, was translated into English, Swedish, and German during his lifetime and was the standard book on the subject during the eighteenth century. This background made him the ideal scientist to pursue the study of sympathetic ink.

Little is known about Hellot's early life, but he was born into a fairly prosperous family in Paris during an age of absolutism and flowering of the arts. Just as he was completing his studies in chemistry, Hellot lost most of his inherited fortune because of an economic crisis and bank failures in France, and he took on the editorship of the newspaper *Gazette de France* from 1718 to 1732. During his stint at the newspaper he developed contacts in scientific, governmental, and manufacturing circles that helped his career later.[9]

In 1735 Hellot obtained the position of assistant chemist (*adjoint chimiste*) at the prestigious Royal Academy of Sciences in Paris. His career blossomed. Promotions followed at the Academy until he secured the highest appointment of *pensionnaire chimiste* in 1743. Hellot joined one of the most powerful and significant groups at the Academy, becoming in essence a professional academician. These working academicians attended biweekly meetings, sat on committees, and filled the pages of the *Histoire et Mémoires* with their articles. Even though most of these "professionals" also held teaching positions or directed technical government enterprises, the Academy was their central focus.[10]

As an industrial chemist, Hellot worked closely with governmental technical enterprises. In 1730 he was appointed inspector of manufacture for the government, a position that dovetailed with his research interests in mining and dyeing. By 1740 he was the government's general inspector of dyeing. Over the next several decades his governmental and academy appointments mushroomed until he was named a commissioner of the Academy and chemist at the Vincennes and Sèvres Porcelain Works. These appointments reflected broader governmental and societal interest in harnessing materials available through mining and improving dyeing methods.[11]

Hellot was so devoted to his profession, and pursued it with so much focus, that he did not get married until he was sixty-five years old. In 1750 he married a Miss Denis, with whom he had had a long-term affectionate friendship. He was so happy with her character and spirit that the union lasted until his death from the aftereffects of apoplexy—a stroke—in 1766.[12]

In the summer of 1736, the German artist whom Hellot credited with advancing knowledge of sympathetic ink showed some members of the Royal Academy of Sciences a magical phenomenon: a rose-colored salt turned blue when it was heated. Not only that, but when the dried solution cooled off, it disappeared. The artist had climbed the hills of Saxony in search of the magical rock. He dubbed it *Minera Marchassitae,* based on a general term for ore including bismuth, cobalt, iron, and other minerals. This curiosity enchanted Hellot, who by the next summer had written a paper on the subject for the Academy's journal; he presented a follow-up paper a few months later. Although Hellot was honest and generous in giving this mysterious German artist full credit for the discovery, he never identified the artist by name. Even so, it is clear that Hellot was the first person to tease out the science of this phenomenon and was the first to experiment with it. As a result, cobalt invisible ink began to be referred to as "Hellot's sympathetic ink" by scientists, popularizers, and the public.[13]

Hellot became so fascinated with what he called this "little curiosity," that he experimented assiduously during his leisure time and began to research the history of the development of sympathetic ink along with looking into the subject of bismuth and cobalt. He collected a bibliography of some fifty pages on bismuth, cobalt, and mining and searched for natural philosophers or scientists who had dabbled in the topic. The paper that emerged from his reading and experimentation was the first major scientific paper on the subject. Even though della Porta had written a chapter on the subject of invisible writing for his *Natural Magic* book, it was really only a collection of the ways in which people could, or did, use invisible writing, not a systematic scientific study of the subject. In fact, Hellot's paper, published in the prestigious journal of the French Royal Academy of Sciences, seems to be the only publicly available systematic study with new information until the twentieth century.

Very soon after Hellot's pathbreaking paper was published,

German scientists objected to the fact that the Frenchman got credit for "discovering" this new phenomenon. They claimed that Professor Hermann Friedrich Teichmeyer in Jena, had, in fact, first demonstrated this cobalt sympathetic ink to his students six years earlier, in 1731. And, of course, there was the mysterious "German lady" who had described a similar phenomenon in 1705. Nationalistic pride on the part of the French and the Germans became part of this priority dispute. Apparently, it was not enough for the Germans that Hellot had generously credited an anonymous German artist with calling his attention to the curiosity. They were surely galled when cobalt sympathetic ink became the eponymous Hellot's sympathetic ink in the centuries to follow, an honor implying discovery.[14]

Until Hellot came along, people such as della Porta who had experimented with sympathetic ink observed that some inks developed by heat, others through application of a glutinous material, and others through application of a "liquor" that worked only when paired with a certain chemical. The latter ink that needs a specific developer, known as a reagent, is the most important and secure kind of pairing. It was the one that Robert Boyle used to make an ink "confess its secrets."

But Hellot developed a whole classification system for these three common developers, while adding two new categories of his own: in the first new type, air developed some dyes and in the second one, some inks, like the cobalt, appeared and disappeared.

The common denominator of all the old secret-writing pairings was that once developed, the writing stayed visible. In other words, even when cooled, the writing did not disappear again. And this is what led Hellot to investigate the anonymous German artist's claims more closely. The remarkable feature of the cobalt sympathetic ink was that when the ink developed by heat cooled off, the writing disappeared.[15]

Hellot experimented with different mineral and solution pairings to come up with his sympathetic ink. At the time, a scientist could not obtain a solution of cobalt chloride at the druggist, but rather had to make it. Hellot bought the mineral from various local druggists or accepted samples of cobalt pieces from his scientific friends from different countries. The German artist had insisted that only the cobalt procured from the mines in Schneeberg created the magical effect. Hellot proved him wrong. He then discovered that there were different species of cobalt. He did note that the most beautiful effect appeared with the purest cobalt. Some samples had arsenic, some had sulfur; experimenters removed other minerals to obtain purer cobalt.

Hellot had experimented with various minerals impregnated with cobalt and assumed that bismuth ores in general produced the peculiar nature of the new sympathetic ink. In 1744, however, Johann Gesner proved that it was the cobalt contained in bismuth ores that was the active agent in producing the ink.[16]

Hellot was fascinated by the color changes when he experimented with the sympathetic ink. Depending on what solution he dissolved the salt with and the purity of the cobalt, he obtained different-colored solutions before heating. When he dissolved cobalt in aqua regia (a mixture of nitric and hydrochloric acids), the sympathetic ink was red in solution, and greenish-blue when developed. Depending on the mineral he matched with a solution, he claimed to have produced colors ranging from green-blue and yellow to reddish-purple, crimson, and pink. If he added sea salt to the developed ink, it became blue. He was most enchanted with a lilac color he produced. Soon experimenters developed a blue sympathetic ink from an acetate of cobalt and a green sympathetic ink from muriate (chloride) of cobalt.

Hellot became so caught up in experimenting with the different color possibilities that he finally started thinking about painting landscapes with the inks or dyes. He hoped that he could find a good painter to create a landscape using all the different colors he had made. When the ink was painted on a sheet of paper, he dreamed of turning a winter landscape instantly into spring. He imagined using the blue colors for the sky and the green for the rivers and the trees. When the paper cooled, it would be winter again.[17] It was truly magical, but Hellot never imagined that he would enchant the general public with this changeable landscape for the next two hundred years.

By around 1746, it became very fashionable in Paris to make sympathetic ink. Recipes abounded. Travelers even went so far as to purchase cobalt in Spain and other European countries to make their own ink. One such traveler claimed that Spanish cobalt produced a much livelier green color than the ink made from the cobalt in Saxony (perhaps because the latter contained more impurities). Unlike the gray Saxon mineral, the Spanish cobalt was blue, like melted lead. As a result, Spanish manufacturers simply pulverized the stone when they took it out of the mine, mixed it with a liquid, and painted it on their wares.[18]

In eighteenth-century Paris, changeable landscape fire screens became all the rage. A barren winter landscape with tree trunks and branches was painted on a fire screen with ordinary India ink. The artist then painted a solution of cobalt chloride on the screen to create lush shrubs and greenery. They used acetate of cobalt to paint blue features like the sky. The cobalt was invisible initially, but as soon as the heat from the fire reached the screen, the barren winter landscape magically turned into a verdant green landscape. When the heat source was removed, the landscape became barren winter again. Although cobalt is not as readily available in the twenty-first century as it was in the eighteenth through the early twentieth century, the reader should try out this sympathetic ink; it is truly magical and beautiful.

From Fire Screens to Weather Dolls and Hygrometers

By the nineteenth century women used handheld paper fire screens to shield their faces from intense heat. They applied the same formula for the small fire screens as people had employed in the eighteenth century: India ink to trace a winter landscape—the tree trunks and leafless branches, acetate or nitrate of cobalt for the blue sky and distant mountains, and cobalt chloride for the green shrubs and leaves. Simultaneously, cobalt sympathetic ink had become part of chemical recreations and amusements.[19]

By the late nineteenth century chemists began to realize that the color change depended on the humidity in the air. When the paper turned from blue or green to a rose color, it meant there was high humidity in the air. This led to the creation of cute "weather dolls" and floral arrangements. These devices were often erroneously labeled barometers. However, they measured not barometric pressure but rather humidity in the air. These chemical toys began to be sold as weather indicators. The doll's dress was impregnated with cobalt chloride and changed color from blue to pink when there was an increase in the amount of moisture in the air.[20]

With Hellot's discovery and the production of the magical changeable landscapes, sympathetic ink had become a real wonder of nature, a curiosity that demanded explanation. A number of distinguished chemists, like the Frenchman Pierre Joseph Macquer (1718–84), who succeeded Hellot as director of dyeing, described and promoted Hellot's sympathetic ink in newly created dictionaries or textbooks for chemistry. This meant that the curiosity merited serious scientific attention. Unlike some of his pharmacy-oriented predecessors, Macquer believed in "chemistry for chemistry's sake." In the end, Hellot found the color changes inexplicable scientifically. As a result of this unsolved scientific mystery, the phenomenon captured the attention of chemists as a serious problem. Early experimenters simply at-

The Smoking Man and Miners with Their Rocks,
Kristi Eide, photograph

tributed the color changes to the effect of heat and cold. By the
late eighteenth century a color chemist, Edward Hussey Delavel,
thought that moisture in the air caused the color changes. He
postulated that the salt attracted the moisture when cold and expelled it when heated. This was the closest early scientists came
to matching later explanations: when the salt is heated it loses
water, an anhydrous state; when it is cooled the water returns to
the salt, a hydrous state.[21]

Hellot's sympathetic ink had inspired the creation of fire
screens and weather dolls, but it also entered the literary imagi-

nation. In 1788, Erasmus Darwin (1731–1802), Charles Darwin's grandfather—a large, ugly man who was a practicing physician, natural philosopher, physiologist, inventor, and poet—was inspired to write some vivid rhyming couplets about cobalt sympathetic ink in his evocative poem *Botanic Garden*, an ode to the goddess of botany:

> Thus with Hermetic Art the ADEPT combines
> The royal acid with cobaltic mines;
> Marks with quick pen, in lines unseen portrayed,
> The blushing mead, green dell, and dusky glade;
> Shades with pellucid clouds the tintless field,
> And all the future Group exists conceal'd;
> Till waked by fire the dawning tablet glows,
> Green springs the herb, the purple floret blows,
> Hills vales and woods in bright succession rise,
> And all the living landscape charms his eye.[22]

In compact form, Darwin captures the wonder of the changeable cobalt landscapes and inspires the reader's imagination. Even though Jean Hellot, the popularizer of cobalt sympathetic ink, was a positivistic Enlightenment scientist, Darwin uses an "adept," an alchemist, as the creator of the changeable landscape. With his magical power, the adept transforms a dead invisible landscape into a vibrant "living landscape" that "charms his eye."

The magician alchemist combines "royal acid" (aqua regia) with the cobalt, dips his pen into the solution, and sketches out the still-invisible scene—a "blushing" meadow, a small green valley, a shadow opening in the forest. Then he "shades" the "tintless" green field's horizon with "pellucid clouds." Like a morning sunrise, the scene is awakened by fire, and the "dawning tablet glows" with green herbs and purple florets. Hills, valleys, and woods become alive and cast a spell on the viewer.

These stanzas were part of the first canto's section devoted to the effect of warmth, electricity, and light on buds. Darwin argued that electricity accelerated germination. He described the buds and bulbs of plants as "buds imprison'd, or in bulbs intomb'd" in winter before heat and light make them blossom. Similarly, in the man-made world, when heat was applied to the cobalt-painted winter landscape it flowered into a verdant green spring landscape. The changeable landscape fire screen mimicked the way nature operated.

The Botanic Garden is a book-length poem in two parts— *The Economy of Vegetation* and *The Loves of the Plants.* More than a poem, it popularized science by stimulating its reader's interest in the natural world with its vivid metaphors and enthusiasm for nature. Darwin's purpose was to "inlist Imagination under the banner of science." The second part, *The Loves of the Plants,* became popular among the public because it anthropomorphized plants and used sexual language. The sympathetic ink stanzas were embedded in the first part, *The Economy of Vegetation,* a poem that celebrated scientific discovery and technological innovation. Much of the poem describes mining and minerals. It also had an evolutionary theme and depicted scientific and social progress as part of evolution and turned inventors into heroes or geniuses.[23]

The poem is annotated. In notes to the sympathetic-ink stanzas, Darwin describes the process by which zaffre (a blue pigment made from cobalt oxide, used to color stained glass), dissolved in aqua regia and water, produces a "fine green-blue"; if zaffre or regelus of cobalt (another oxide of cobalt) was dissolved in nitre (nitric acid), or aqua fortis, it produced a reddish color. But what he thought was "more wonderful" was that both lost their color when they were removed from the fire. Zaffre— unlike pure cobalt chloride—was readily available at the drugstore from Darwin's time well into the twentieth century. Then it disappeared from public availability—presumably because of cancer risks when it is used carelessly.[24]

Erasmus Darwin was a member of the Lunar Society (1765–1813), a discussion group of British natural philosophers and industrialists that met on evenings lit by the full moon so that they could walk home safely. Several notable members included James Watt, inventor of the steam engine, Joseph Priestley, the experimental chemist, and James Keir, a pioneer in the chemical industry. Darwin refers to Keir's *Chemical Dictionary* in the annotation of his poem. Keir had translated Pierre Macquer's *Dictionary of Chemistry,* which included an entry on Hellot's sympathetic ink, and it is likely this is how the group heard about the chemical underpinnings of the magical changeable landscapes. These translations, along with word of mouth about the wonderful color changes, helped the decorated fire screens make their way from France to Britain.

Erasmus Darwin was one of the first British popularizers of science to appear in print during the eighteenth century. During the last two decades of that century of Enlightenment, many popularizations of science followed in Britain, France, and Germany. Late-eighteenth-century Britain had become home to a second scientific revolution that created an era of romantic science and produced what Richard Holmes has called an Age of Wonder. Other British scientists were also enchanted by the wonder of science and wrote poetry extolling its virtues.[25] The popularization of science soon led to a magical age for invisible ink.

Meanwhile, as Hellot's sympathetic ink enchanted Europe, revolutionary storm clouds began to gather across the Atlantic in colonial America. As the Revolutionary War progressed, George Washington desperately started searching for a secure way to communicate with his agents.

Revolutionary Ink

5

IN 1807 SIR JAMES JAY (1732–1815) petitioned Congress to compensate him for inventing a secret-ink system during the American Revolution. Because of the *"peculiar* nature" of the "services rendered," Jay had not asked for, or received, payment during the war. Since there were no formal intelligence bureaucracies at the time, and this was an unprecedented request, Congress was unsure how to handle the request. It was not until more than a century later that American intelligence developed formal units for developing secret communication tradecraft. As a result, Jay's sympathetic ink was lost to history, and detective work is needed to determine what the formula might have been. Solving the enigma of the composition of the secret ink used during the Revolutionary War—by both sides—is an attractive challenge, but it is not just an end in itself. For it turns out that General George Washington was enamored with this new technique and with espionage in general. In fact, along with deception, good secret communication was a major factor in helping him win the war.[1]

James Jay was a noted physician who had sailed from New York City, where he was born, to Great Britain to receive his medical education and an M.D. from the University of Edinburgh, one of the world's leading medical schools in the eigh-

Portrait of James Jay

teenth century. But his last name might not be familiar except
for the fact that his younger brother, John Jay, was a Founding Father of the United States and the first chief justice of the
Supreme Court. Upon his return to New York from London
in 1756, James Jay practiced medicine, but managed to alienate
most of his patients because he was "haughty, proud, overbear-
ing, supercilious, pedantic, vain and ambitious," not to mention that he charged excessive fees. Soon his only patients were
relatives. As a result of the rocky start to his medical career in
New York, he returned to England to obtain endowments for
King's College (now Columbia University) and the University
of Pennsylvania. King George III knighted him in 1763. After
prevailing in a decadelong legal battle over the monies for the
colleges, he was back in England when he began to hear the
rumblings of revolution in America.[2]

Despite his time in England, James Jay was patriotic to the revolutionary cause, at least during the early years. While he was in England, he developed a sympathetic ink to communicate secret military information back to his brother in America. Long after the war he prided himself on informing the Continental Congress of the "determination of the British Ministry to reduce the Colonies to unconditional submission." In addition, he told Benjamin Franklin and Silas Deane in Paris about a planned British invasion from Canada.[3] But he never disclosed the composition of the secret ink.

James Jay also provided his younger brother with "considerable quantities" of sympathetic ink to communicate with him and other patriots. One of the first recipients of this magical fluid was Silas Deane (1737–89), a secret agent for the Continental Congress. John Jay had provided him with his brother's invisible ink as Deane boarded a ship to Paris in April 1776. With his easy smile and ready laugh, the gregarious Yale College–educated Deane was a good choice to send to France. His assignment was to buy arms, clothes, and ammunition for the Continental Army under the cover of a merchant named Thomas Jones, while feeling out France as an ally should the colonies declare their independence. Since John Jay and Robert Morris had been appointed by the Continental Congress to its "Secret Committee of Correspondence," they communicated with Deane, who became the first American agent to use invisible ink.[4]

Deane had served as a delegate from Connecticut to the Continental Congress between 1774 and 1776, but his appointment as secret emissary to France seems puzzling. After a brief stint as a lawyer, Deane had earned a living as a merchant and shopkeeper, had hardly left Connecticut, and didn't even know French. Earlier he taught school, where one of his pupils was the future double agent Edward Bancroft, a contact that would come back to haunt Deane later. It was, in fact, precisely because

Deane didn't seem like a secret agent that Benjamin Franklin chose him for this mission.[5]

By June 1776, Deane had arrived in Paris. Since his mission was confidential and he was told to pretend that he was in Paris to see the famous city's sights, he wrote mounds of invisible-ink letters to John Jay and Robert Morris, the American Revolution's financier. He would take a large sheet of paper and write an ordinary letter on the top of the sheet and a secret invisible letter on the bottom.

In a missive to Robert Morris, the ordinary letter told of his arrival and mentioned that he was sick from the trip, but below that message he wrote about an impending shipment of "clothing for twenty thousand men" and arms and equipment for the war. He also told Morris of the diplomatic situation in Europe, advising him to secure an alliance with Spain and to increase the size of the navy; he also predicted that general war was likely in Europe. Morris didn't have the secret developer to read the letter, so he sent it to John Jay, who applied the reagent and then transcribed the letters for him.[6]

Robert Morris asked John Jay to develop and transcribe "the Invisible part" of many of Deane's letters. Morris knew that the two of them had an arrangement to use "a mode of writing that would be invisible to the rest of the World," but Deane had promised to ask Jay to communicate the secret to him as well. Jay never did reveal the secret to Morris, even though he acknowledged that he "communicated to" Deane a "Mode of invisible writing unknown to any but the Inventor & myself" and claimed "the enclosed letter will explain it." But there was no enclosure to Morris telling him more about the invisible ink system, because John Jay really did not want to share this secret. Later he promised to tell Deane in person. But instead of disclosing the secret, Jay became the central clearinghouse for Deane's secret letters.[7] Even if John Jay knew that the system required the invisible ink and a reagent to develop it, there is no evidence that he knew the actual chemical composition of the two substances.

Morris acknowledged receipt of the developed Deane letters in cryptic terms. He spoke of "Timothy Jones" (Deane's code name) as "a very agreeable, entertaining man . . . one would not judge so from anything contained in his cold insipid letter of the 17 Sept., unless you take pains to find the concealed beauties therein; the cursory observations of a sea captain would never *discover* them, but transferred from his hand to the penetrating eye of a *Jay*, the diamonds stand confessed at once."[8] Of course, the "concealed beauties" and the diamonds that "stand confessed at once" are the invisible writings that have been developed. Just as Robert Boyle in the seventeenth century wrote of invisible ink confessing its secrets when developed, so too did eighteenth-century statesmen use this wonderful image.

The Culper Ring

Meanwhile, there was a flurry of activity in the homeland. The British had occupied New York City and made it their headquarters. General George Washington needed information about their plans and military strength and asked his intelligence chief, Major Benjamin Tallmadge, to find an intelligent person in the city to act as a messenger. Devastated by the hanging death of his friend the patriot Nathan Hale, Tallmadge was determined to develop a secure spy ring under the strictest of secrecy. Even Washington did not know the true identity of the spies of what became known as the Culper Ring. Their identities were so secret that not until 1930 did a Long Island collector and historian find evidence linking the code name Samuel Culper Jr. to an obscure merchant.[9]

Abraham Woodhull, a trusted friend of Tallmadge from his

George Washington at Morristown,
John Ward Dunsmore

hometown of Setauket, Long Island, was a modest farmer who
traded vegetables and livestock from his farm for luxury goods
like German mustard, Spanish olives, and Scottish smoked
salmon available in New York City. Manhattan had become an
island fortress. In a sense, New York City then was like West
Berlin during the Cold War, an island surrounded by an enemy
that spoke the same language. Though the British controlled
access to New York City, they became dependent on local
farmers for basic goods because Washington had power over
the surrounding rural towns. The British ended up importing
more than they exported. Since Woodhull was adept at smuggling goods, Tallmadge thought he would be equally good at

smuggling secrets. When he was recruited in the fall of 1778 he took the oath of loyalty and was provided with the code name Samuel Culper Sr.[10]

Woodhull had the plausible excuse of visiting his sister, who, with her husband, ran the Underhill Boardinghouse in New York City. While he was in the city, Woodhull, hollow-chested and thin, wrapped in an oversized overcoat, mingled with the British occupation forces, rubbing shoulders with crowds in marketplaces and hanging out at coffeehouses, all the while picking up gossip from incautious military officers. Thus he gathered information about troop movements, shipping, and supplies. But he was getting cold feet, living in constant fear of discovery.[11] He trembled every time he went through the British checkpoint and had to show his passport. He begged his chiefs to destroy his letters instantly after reading them. Washington was also concerned about letters falling into enemy hands and wished there was a "mode of correspondence" that existed whereby Culper would have "nothing to fear."[12] Washington was also eager to receive information in a timelier fashion.

The problem was that their intelligence reports were written in black ink. If Woodhull was stopped and searched and the documents intercepted, it would spell disaster for Washington. In fact, Benjamin Tallmadge himself was the victim of such a search when he was riding his horse on the way to see George Washington.

The close call meant it was time to increase security. John Jay had recently told Washington about the magical white ink, which his brother had developed and which he himself had used for several years. Washington, who had been looking for a way to make correspondence more secure, was enthusiastic about the "new mode of correspondence." Even though it took almost six months for Washington to receive the magical fluid from Jay,[13] he began to experiment with it and marveled that Jay's new "sympathetic ink" was impervious to heat: "Fire which will bring out lime juice, milk and other things of this kind to light, has no effect on it. A letter upon trivial matters of business, written in common ink, may be fitted with important intelligence which cannot be discovered without the counterpart."[14] Unlike simple invisible inks like lime juice, this was a more sophisticated kind that required a chemical developer, or reagent, to develop the agent. Washington soon dubbed this magical ink "sympathetic stain."

But even when Tallmadge provided Woodhull with Jay's secret ink, it did little to make him less nervous. One day he was sitting alone in the attic room of his sister's boardinghouse preparing an invisible-ink letter to Tallmadge. All of a sudden the door burst open, and he leapt to his feet and knocked over the table with the precious invisible ink. It turned out the intruder was his sister, not British boarders set on making an arrest. As a result of this "excessive fright," he asked again to be relieved of his duties.[15]

Because of these fears, Tallmadge and Washington were looking for an agent in place in New York City who could pass on information. A recruitment opportunity came when "Samuel Culper" confided in a good friend about his new mission and recruited him into the growing spy ring. Woodhull's confidant was Robert Townsend, a patriotic New York City merchant and Quaker. He had an excellent cover because he could travel back and forth between the city and Long Island doing business. To add to his cover he pretended to become a Tory by joining the militia and writing antipatriot articles for the *Royal Gazette*, the Tories' mouthpiece. He also became partners with the *Royal Gazette*'s printer, James Rivington, by opening a coffee shop, where he heard much gossip from British officers.[16]

Townsend, now code-named Samuel Culper Jr., gathered intelligence in New York City, composed letters in invisible ink and code, and passed them on to Austin Roe, a courier hired to work with the Culper Ring. Roe owned a tavern in Setauket and rode his wagon into New York City to collect goods for his tavern. Now his new goods were entirely of a different kind. When Roe returned to Setauket, he passed the intelligence on to Woodhull, either directly or through a dead drop in a container buried in a field on his farm.

Once Woodhull had the letters, he would examine his neighbor's clothesline. If there was a black petticoat hanging on the line, it meant that Caleb Brewster, a whale boatman and member of the Culper Ring, had arrived and was hiding in one of the coves, as indicated by the number of handkerchiefs on the clothesline. At night he would sail past British guards and transport the report from Setauket across the Long Island Sound to the town of Fairfield, on the Connecticut coast. From Fairfield a courier on a fast horse took the report to Tallmadge, who would view it and send it on to Washington's headquarters via yet another horse-riding courier.[17] Through this new relay method, Tallmadge and Washington learned that the British were building transports in the middle of winter and initiating raids on Connecticut.

Later in the summer, Washington told Benjamin Tallmadge: "All the white Ink I now have—indeed all that there is any prospect of getting soon—is sent in phial No. 1 by col. Webb.—The liquid in No. 2 is the counterpart, and brings to light what is wrote by the first, by wetting the paper with a fine hair brush— These you will send to C—Jnr.[?] as soon as possible, and I beg that no mention may *ever* be made of your having received such liquids from me, or any one else."[18]

It was also in the summer of 1779 that Washington learned that the British governor of New York City was probably using the same kind of invisible ink as he was. But this did not deter Washington from using it; he was hoping the enemy would never even know that his spies were using it at all: "I am informed that Governor Tryon has a preparation of the same kind, or something similar to it, which may lead to a detection if it is ever known that a matter of the sort has passed through my hands."[19] Though it was unfortunate for Washington, the fact that the British were using the same kind of invisible ink is a clue as we search for its composition.

Washington thought very carefully about the mechanics of security and secret communication. As an extra precaution, everyone in the Culper Ring also used a code in addition to invisible ink and received a number or pseudonym: Tallmadge became John Bolton and the number 721, Washington was 711, New York City was replaced by 727, and of course Townsend and Woodhull were the Culpers.

In the fall of 1779, Townsend/Culper Jr. considered leaving his real employment as a merchant to conduct espionage full-time. Washington didn't think this was a good idea because carrying on with his regular employment offered more security and cover. He also advised Townsend to write secret messages on the "blank leaves of a pamphlet . . . on the first second & c. pages of a common pocket book — on the blank leaves at each end of registers for the year to render his communications less exposed to detection. Apparently this secret ink worked better on quality paper. This way of communicating would also relieve Woodhull/Culper Sr.'s fears.[20]

About a year after the Culpers had started receiving the magic ink, Washington reported to Jay that "the liquid . . . is exhausted." Since Woodhull was so fearful of discovery, he tended to hoard the ink. Washington was quite irritated about this and told Tallmadge that "what I have sent him at different times would have wrote fifty times what I have recd. from him."[21] As a result of this new shortage, Washington asked James Jay to provide him with more of the "very useful" substance. He even gave Jay the authority to procure the chemicals under Washington's name from a hospital if he did not have the "necessary ingredients."[22]

James Jay then sent him "the medicine" (as they began to call it) "in a little box." This was all he had left from the supply he had brought back with him from Europe. Although he had the "principal ingredients for the composition" with him, he did not have a place to prepare the ink, for "the composition requires some assistance from Chemistry; and our house is so small . . . that there is not a corner left where a little brick furnace . . . can be placed." He thought a "log hut" would suffice, but he didn't have the materials to build one. Since he didn't want Washington to think he was declining the "undertaking," he asked for the provision of workmen and supplies so he could "have the satisfaction" of sending the general as much of the ink as he wanted to use "freely . . . without the apprehension of future want." Washington was more than glad to provide assistance in "erecting a small Elaboratory, from which" he hoped Jay would "derive improvement and amusement and the public some advantage."[23]

In the meantime, the ring was successfully transmitting important information about British troop movements and fortification and plans in New York and the region around it.

James Jay's Invisible Ink

Even though George Washington worked closely with James Jay in order to procure the valued "sympathetic stain," he never knew what it was, nor did John Jay have a clue. In 1935, Dr. Lodewyk Bendikson (1875–1953), a former physician and innovative photographer at the Huntington Library in San Marino, California, applied his new ultraviolet photography method to uncovering obliterated passages and invisible ink in letters to John Jay. He

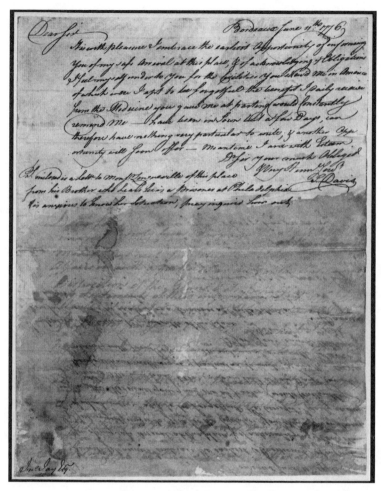

Silas Deane letter, June 11, 1776

focused on several invisible-ink letters Silas Deane wrote to Jay in 1776. Remarkably, Bendikson was allowed to use the original letters sent to him by Dr. Frank Monaghan, Jay's biographer, to reveal the obliterated passages and invisible ink. These letters have now disappeared, though they were available in the rare-

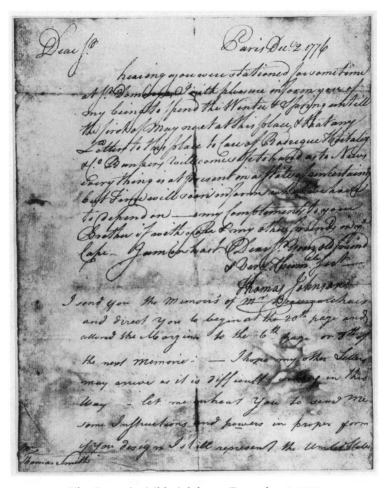

Silas Deane invisible-ink letter, December 2, 1776

documents catalogue of the bookseller Kenneth Rendell in the 1970s. Once Bendikson completed his evaluation of the secret-ink letters, he returned the letters to Monaghan, who discussed them with Rendell in the 1960s. As this book was going to press, I turned up one more invisible-ink letter in the collection of Joe

Lodewyk Bendikson at his camera

Rubinfine, who claims not to know what happened to the rest of the letters. Bendikson's descriptions and photographs of the letters, however, remain.[24]

During the 1930s Bendikson started applying new methods from modern science and technology. He used analytical chemistry and ultraviolet and infrared photography to reveal the invisible in old documents with the magical rays. When Bendikson turned his ultraviolet camera to Deane's December 2, 1776, letter, it "leaped at him as though written in fire." Even though the letter was 150 years old, the chemical salts still glowed, or fluoresced, with the application of the ultraviolet rays. Bendikson then photographed the letter.

Deane, who used the code name Thomas Johnson, wrote John

Jay a visible letter on the top half and used the bottom half for the invisible missive. This seems foolhardy from our perspective, but in the eighteenth century such letters were not suspicious. More than anything else, the hidden text reveals Deane's anxious state because of his failed attempts to find intelligence as John Jay's "plenipotentiary."[25]

Bendikson was also curious about the composition of this secret ink. He hired George D. van Arsdale, a Pasadena, California, chemical engineer, to help him out. Without doing any research into the available chemical combinations in the eighteenth century, van Arsdale found out that James Jay's invisible ink consisted of a solution of tannic acid, or gallnuts. John Jay then sponged the paper with ferrous sulfate to reveal dark letters, easy to see against the light paper.[26]

British Secret Ink

Although George Washington was aware that the British used invisible ink, he didn't know that it was the British who wrote the first invisible-ink letter during the American Revolution. In fact, British spies anticipated Washington and Jay by several years.

In 1950, Professor Sanborn Brown, a physicist at the Massachusetts Institute of Technology who had worked on secret inks at the U.S. Office of Censorship during World War II, discovered what he considered the first secret-ink letter of the American Revolution. It was no longer invisible, having been developed either by the intended recipient or at some time since. Brown did some sleuthing to determine that the letter with an obliterated signature was written by Benjamin Thompson, also known as Count Rumford, considered "one of the most brilliant scientists of the era."[27]

Thompson was driven from his hometown in New Hamp-

Benjamin Thompson, invisible-ink letter, first page

shire in December 1774 because angry citizens there suspected
he was a Tory. As a result, he ended up in Woburn, Massachusetts (indicated in the dateline of the letter), and established headquarters there until he openly joined the British in Boston in
October 1775. Meanwhile, he needed to communicate secretly

with the British because he disclosed sensitive movements and plans of the "Rebel Army." On May 6, 1775, he reported to King George that the rebel army had grown to thirty thousand "effective men" and planned to attack Boston. He also informed the king that Congress intended to execute its plan for "*Independence* at all adventures." Clearly, Thompson couldn't communicate this information in regular ink, because if the patriots read the letter, he would be caught and hanged.

Brown, along with Elbridge W. Stein, an examiner of questionable documents, determined that the composition of the ink was tanno-gallanic acid, and the reagent was iron sulfate; this combination, as we have seen, was known by the ancient Greeks and della Porta, and was still being used during the Enlightenment. For a long time, before the arsenic formula and cobalt, it was really the only chemical preparation available and the best one requiring a specific developer. Brown and Stein examined the document with a very low voltage X-ray machine and ruled out lead and bismuth inks. They found this result to favor the gallnut combination. Since the letter had already been developed with the iron sulfate, it became a dark purple under ultraviolet light. The gallnut solution also becomes visible under ultraviolet light without a developer.[28] If all these tests were accurate, it suggests that the British and Americans used the same kind of ink, as General Washington surmised.

In addition to secret ink that was impervious to heat and other easy detection methods, the British used old-fashioned lime or lemon juice and other substances that heat could develop. Major John André, head of British intelligence in New York, told agents to mark the top right corner of their letters with an "F" if they used invisible ink developed by fire or heat and an "A" for those that needed an acid.[29]

Aside from low security, the primitive methods led to many mishaps. When Jonathan Odell, a Loyalist clergyman, noticed the "private mark agreed on," he "assay'd" his secret-ink letter "by the Fire, when to my inexpressible vexation, I found that the paper, having by some accident got damp on the way, had spread the Solution in such manner as to make the writing all one indistinguishable Blott." He also noted that "toasted paper becomes too brittle to bear folding." As a result, he transcribed most invisible-ink letters and discarded the originals, much to the annoyance of historians.[30] Other spies also told their confederates to "toast" their letters to reveal the invisible part with fire.

But the plot thickens when we bring in the British and their use of invisible ink during the Revolutionary War. One of the greatest mysteries of the period was whether or not Dr. Edward Bancroft, a British spy, physician, and scientist, poisoned Silas Deane, causing his mysterious death as he was boarding a ship in France to take him back to America in 1789.

When Deane traveled to France, Benjamin Franklin gave him six names to look up. Deane did so gladly and immediately, because he did not know anyone and did not know French. One of his first contacts was with Bancroft, whom he had briefly tutored when Bancroft was a teenager in Connecticut. Franklin warmly recommended Bancroft as a friend of the American rebels. He was also impressed with Bancroft's scientific discoveries concerning inks, dyes, and electric eels. Even though Bancroft had only been a brief contact initially for Deane when he was in Paris, their friendship deepened.[31] But there was one thing Deane did not know.

The day after Bancroft returned to England from Paris, he became a spy for the British. By August 1776, Bancroft had related everything he had learned from Silas Deane in Paris to the British secret service. They knew before the Continental Congress of Deane's new friendships, shipping details, and supplies sent to America. And they didn't even have to read his secret-ink letters.

James Jay's Quest

James Jay clearly helped the patriot cause by providing Washington's spies with secret ink. With his "Elaboratory," Silas Deane and George Washington were dependent on him as a supplier of the sympathetic stain. There is no doubt that Washington appreciated Jay's efforts. He even thought it helped him win the war. But for some reason he never paid Jay for his efforts.

Long after the Revolutionary War was over, Jay sought compensation first from President Thomas Jefferson and then from Congress. Initially, Jay asked Jefferson for $20,000. Presumably, Jefferson referred him to Congress, because he prepared a "memorial" on this issue for them. Jay declared that this secret mode of communication had been useful during the American Revolution and could be of public utility in other times as well. But the matter of how to handle secret ink nevertheless became quite controversial in congressional hearings. A lively debate ensued. "Much wit was displayed" in describing different ways of keeping secrets and the "futility of all." After all, they quipped, the House seemed to know Congress's secrets before they were disclosed.[32]

Those in support of compensating Jay recognized the invention's importance for the Revolution, while those opposed thought it was "absurd to vote away money, for a thing they did not and could not understand." Opponents were also worried that the secret might be disclosed to foreign governments.[33]

The House resolved that "it shall be lawful" for the president to buy the system invented by Jay "provided" he thought it would be of "public utility and importance to possess" it. The House ended up passing the petition by a one-vote margin on November 21, 1807. On March 2, 1808, Jay again petitioned the House, this time, "praying the liquidation and settlement of a claim against the United States, for moneys advanced, and services rendered, of an important and secret nature, during the Revolutionary war with Great Britain." This time the House did not approve or disapprove, instead voting to postpone consideration of the matter "indefinitely." The Senate did not consider it again until July 7, 1813. Even though Congress acknowledged that Jay had "imparted a plan of secret correspondence, which proved to be of great importance in the course of the war," he died two years later without being reimbursed for his efforts.[34]

Magic

IN 1922, THE PORTER Chemcraft Company advised new owners of their chemistry sets how to set up tricks and prepare a magic show to "amaze" their friends. Porter chemists suggested dressing up as alchemists, who had in days past been seen as "wizards" or "magicians" who brought about wonderful color changes in materials. The chemical magic tricks listed in the manual used simple chemical reactions to produce "beautiful color changes, thick clouds of smoke without fire, diabolical odors," and "invisible inks."[1]

For many people, invisible ink conjures up images of childhood magic and fun, but most of us don't know how we came to view it in this way. And even when we have dug up the past and discovered the origins of the notion of magical sympathetic ink, there is a missing link in the chain of events leading to this magical view. On this count, the implication in the Chemcraft Company brochure is wrong. Even though alchemists brought about wonderful color changes in materials, they were not the people responsible for propelling invisible ink into the world of stage magic.

The story, of course, begins with Jean Hellot's discovery of the appearing and disappearing ink, but it does not develop until an unexpected entanglement occurs in the eighteenth century, an entanglement that could not have happened in the seventeenth. During the last decade of the Age of Enlightenment, the new secular stage magic[2] movement coincided with a revival of natural magic and the scientific recreation movement to create a lively era of popular science. It was against the backdrop of this witches' brew that magical sympathetic ink flourished. And the missing link that brought magical ink into the twentieth century was the chemistry set.

Popular Science

One of the first steps that brought about an age of popular science in the eighteenth century was a change in the venue where science was done. While in the seventeenth century science usually took place at the Royal Academies of Sciences, at the universities, at the king's court, or in the private quarters of scientists, by the early eighteenth century several new sites for science had emerged. One of these new places was the coffeehouse. In 1739, London alone was home to more than five hundred cafés, many of which offered natural philosophical lectures. Natural philosophers adorned their lecture courses with experiments and demonstrations based on Newtonianism and the new mechanical philosophy. Their audiences included a varied group of eager listeners: some customers stopped in for a cup of coffee and scientific entertainment, while others came in to read the newspapers. Many customers were financiers and merchants who exchanged business information and listened to lectures. Philosophes like Rousseau, Diderot, d'Alembert, and the American scientist and statesman Benjamin Franklin were also regulars at the cafés. Voltaire even had a favorite table and chair at the Procope and had a reputation for drinking dozens of cups of coffee.[3]

Along with new homes for scientific lectures came a transformed popular science legacy from the Renaissance princes' courts, the curiosity cabinet (*Wunderkammer*). During the Renaissance, wealthy supporters of science collected bizarre or exotic artifacts from the New World; they identified some as exotic seashells or stuffed crocodiles, and they misidentified others as unicorn horns or giants' thighbones. Collectors also included ornate scientific instruments like microscopes and telescopes in their collections. The rooms were crammed with exotic gems, automata, fossils, antique coins, exotic paradise birds, and animal and human monsters. By the late eighteenth century and early nineteenth century, these private curiosity cabinets had been transformed into commercial science museums and scientific exhibitions.[4] Wonders were now considered vulgar and for the masses.

As curiosity cabinets turned into museums, other institutions like the Royal Institution, the Royal Polytechnical Institution, and the Colosseum in England and the American Philosophical Society in America helped spread popular science. At the Colosseum, a chemist, William Leithead, was in charge of the Gallery of Natural Magic.[5]

In France science had spread from private laboratories and elite royal societies to the popular press and urban culture. Philosophical salons, cafés, and musées dotted the streets of Paris, in particular the Boulevard du Temple, and these new institutions for popular science came alive with scientific lectures, experiments, and demonstrations of science. A French writer, Louis-Sébastien Mercier, observed in his *Panorama of Paris,* "Everywhere science calls out to you and says, 'Look.'"[6]

But the type of science that called out to pedestrians on the streets of Paris was not your typical academic science. It was alternative science, sometimes even pseudoscience, but always popular science. Franz Mesmer mesmerized Parisians with his experiments on animal magnetism, Jean Nollet electrified them with shows in which a charge pulsated through a row of people, one by one, and they were blown away by the ballooning craze started by the Montgolfier brothers.

According to Robert Darnton, during the decade preceding the French Revolution, mesmerism took over French people's imagination. Popular science was on their mind, not the principles of liberty, equality, and fraternity. Literate Frenchmen and women were "intoxicated by the power of science."[7] They were entranced with Mesmer's notion of animal magnetism because it postulated that there was a magnetic fluid inside them that could be manipulated and produce physical effects. He could make their bodies twitch or become paralyzed; he could make them sleep, induce hysteria or coma, or even cure their diseases. Not only was mesmerism intoxicating, but it was also magical. It was in this context of popular science that stage magic grew and flourished.

Natural Magic Revisited

As science became part of public culture, it provided entertainment as well as general education. This led to the emergence of scientist-magicians like Henri Decremps (1746–1826), a self-styled "professor and demonstrator of amusing physics," and Joseph Pinetti (1750–1800), known as the Professor of Natural Magic. They both toured Paris and London performing "amusing physics and various entertaining experiments." But the "natural magician" Pinetti was no della Porta. Although he claimed to have been a professor of mathematics in Rome before moving to Paris, Pinetti was a stage magician, pure and simple. Decremps took it upon himself to unmask charlatans, and a book he wrote exposing and explaining Pinetti's experiments as mere tricks ruined the career of the leading magician in Paris. His displays were no longer a mystery, and his formerly huge audience disappeared. The discrediting of frauds and charlatans soon became a pastime of many Enlightenment scientists, who sought to eradicate superstition from natural science.[8]

Decremps also included sympathetic inks among the secrets he exposed. Now ordinary people at home could choose five kinds of sympathetic inks developed by liquid, air, powder, or fire, and create amusements like the changeable landscapes, the book of fate, and the oracular letters.[9]

Della Porta's notion of natural magic was transformed and extended as the new developments and discoveries in science stimulated an interest in secular stage magic. Della Porta had defined natural magic as "nothing else but the survey of the whole course of nature." He called it "natural magic" because he found the natural world enchanting. The natural magician attempted to uncover the hidden or occult qualities of nature using science as a key to unlock these secrets of nature. It was science, not hocus-pocus or sorcery. By the end of the eighteenth century there was a resurgence of interest in the subject of natural magic. Some natural magicians were conjurors and stage magicians who manipulated reality through sleight of hand, while others were natural scientists enchanted by the natural world.

In addition to the proliferation of new institutions for science, theaters for magic shows, and scientist-magicians, there was an outpouring of new books on "natural magic," on "magic, or the magical power of nature," on "rational recreations," on "philosophical amusements," and on "popular amusements."[10] In a sense, these new natural magic books were updates of sixteenth-century books of secrets. Instead of recipes for the household, these new best-sellers outlined how to perform magic tricks based on the new science.

The new rational recreation movement began in France and quickly spread to England and Germany. Edmé-Gilles Guyot (1706–86), a French physician, cartographer, inventor, and science popularizer, led the pack with his application of the new science to magic. He created apparatuses and scientific instruments for magicians and pieces for clients' curiosity cabinets. The magic lantern, a home theater consisting of a box with a lantern and moving slides inside that projected stories onto a screen—like the image of a storm at sea, for example—was his best-known entertainment. Guyot published his science-supported magic in 1769 in his *New Recreations in Physics and Mathematics*. He built on, and transformed, a late-seventeenth-century trend of packaging magic tricks for science education popularized by Jacques Ozanam (1640–1717), the author of *Mathematical Recreations* (1694).[11]

Guyot was widely admired—and almost as widely imitated —in France, England, and Germany. It seemed as if everyone copied or adapted his work, sometimes with attribution, sometimes without. Pirated editions proliferated. A William Hooper (fl. 1770s) in England based his 1774 compilation *Rational Recreations* on the works of della Porta, Ozanam, and especially Guyot, with some modifications. Hooper aptly captured the goals of rational recreations: "to render useful learning, not dull, [and] tedious ... but facile ... delightfully alluring, captivating."[12]

German natural magic books also began to adapt Guyot's recreations and sympathetic-ink instructions by combining elements of popular science with demonstrations of popular amusements. These new natural magic books contained less and less on experimental physics and more and more on spectacular physical effects. They epitomized Enlightenment ideals by combating superstition of any kind. Tricks were based on science, not on devil's work or sorcery. Just as Decremps had exposed Pinetti as a charlatan, so too did the German natural scientists combat deceivers by exposing the science behind their tricks. They wanted to enlighten readers by showing them that abracadabra was just plain nonsense.

Johann Thenn, the German translator of Guyot's four-volume *Recreations,* introduced a number of changes in his rendering of the work. Most notably, the sympathetic-ink section is considerably longer and more detailed than Guyot's. It includes a dizzying array of amusements based on six kinds of sympathetic inks. It also includes copperplate diagrams absent in the original French. The German natural magic authors relied on the German translation of Guyot, not the original French, when they compiled their derivative natural magic volumes.

Books on natural magic by Johann Christian Wiegleb, Johann Samuel Halle, and Christlieb Benedict Funk, all published between 1779 and 1783, began to emphasize instruction in showmen's tricks for sympathetic ink. By adapting, modifying, and adding to Guyot's "occult writing," these books built on, and transformed, the "rational recreations" movement that entertained an emerging educated class during the Enlightenment. Along with experiments and demonstrations using electricity, magnetism, and optics, their pages were filled with card tricks, numerology, and the book of secret style recipes for preserving sour cherries and instructions for making magic mirrors.[13]

One stunning example of a monumental series on natural magic was Johann Christian Wiegleb's *Natural Magic made from Amusements and Useful Magic Tricks,* first published in 1779. With the later help of Gottfried Erich Rosenthal, Wiegleb's book went through numerous editions, and by 1800 it had mushroomed into twenty volumes. Wiegleb (1732–1800) was one of Enlightenment Germany's most knowledgeable and respected apothecary-chemists. He was known as an "apothecary from Langensalza" in Thuringia, but he was much more than a pharmacist. The son of a lawyer, he was destined for the clergy until he discovered pharmacy and decided to become an apprentice at the Marien-Apotheke in Dresden. His six-year apprenticeship was a sore disappointment, and Wiegleb felt as if the journeymen treated all the apprenticed boys like "slaves." All they did was "drill" and learn things by rote with no rhyme or reason. As a result, Wiegleb had to educate himself in chemistry through a broad program of reading in the general chemistry literature he found at the pharmacy's small library.[14]

Even though the pharmacy education was a disappointment, Wiegleb himself became a journeyman and returned to his hometown of Langensalza to administer a pharmacy, eventually opening his own shop. After he established himself, he began to study chemistry more seriously. He experimented in his pharmacy laboratory, wrote papers and books in his home library, and published in the best journals and with the best publishers. By the late 1780s Wiegleb had become one of Germany's most prominent chemists in a newly formed professional community of chemists. In 1780 his portrait—displaying a serious-looking, long-faced enlightenment scientist with a short wig and prominent nose—appeared on the frontispiece of Germany's premier reviewing journal. This was the first time a chemist's portrait had appeared in a journal. Karl Hufbauer notes, "Wiegleb was a small-town apothecary whose only claim to fame was as a chemist."[15]

Actually, his claim to fame was not only as a chemist, though he was generally known and respected in the eighteenth-century community of scientists as the apothecary from the provincial town of Langensalza. As the Englishman C. R. Hopson noted in 1789: "apothecaries in Germany . . . are . . . possessed of more chemical knowledge than any other body of men in that country."[16] So while Wiegleb was a well-respected and knowledgeable chemist during his time, his more lasting claim to fame was his natural magic book.

Wiegleb did not come to the topic of natural magic on his own. His publisher, Christoph Friedrich Nicolai, asked him to revise and expand a popular natural magic book written by the obscure medical doctor Johannes Nicolaus Martius. Nicolai had wanted to update the book by incorporating the dazzling new results in eighteenth-century chemistry and physics. Originally a dissertation in Latin focusing on medical cures, Martius's tome was published in German in 1717, and by 1751 the Martius natural magic book had gone through numerous printings and editions. He was interested in applying natural magic to curing diseases, and the phrase "natural magic" drew a large and curious readership.

In Wiegleb's hands the old-fashioned natural magic book was totally transformed. It had an introduction on the nature of the magic and new chapters on the new science, as well as extensive chapters on sympathetic ink, among other chemical magic and amusements. Most important, Wiegleb made his abhorrence of superstition clear. He hoped that the "devil's work on earth" had been totally destroyed through the true laws of nature and mathematics. He was amazed that a mere fifty years before his time people still believed that cobolds and ghosts inhabited nature and that princes handed out rewards for catching them dead or alive. He felt that some charlatans would still try to deceive the gullible through magic. But he made it clear that the new magic had nothing to do with black magic and sorcery. It was natural and white magic. It was an art that brought about events that seemed to go beyond the natural power of the body but could really be explained by natural laws. Most magicians' tricks were based on speed or preparation.[17]

The explosion of interest in natural magic coincided with the emergence of portable chemical cabinets. The German chemist and pharmacist Johann Friedrich August Göttling (1753–1809) built one of the earliest "portable chests of chemistry" in 1791, grandly described as "A Portable Chest of Chemistry or a complete collection of chemical tests for the use of chemists, physicians, mineralogists, metallurgists, scientific artists, manufacturers, farmers and cultivators of natural philosophy." Interestingly, test tubes are never mentioned. Göttling recommended using a wine glass to mix chemicals for color and precipitation reactions that did not require heat. He conceived of this chest as more than just a professional chemist's kit, because he said it could be used for "chemical instruction and amusement." He also recommended consulting the "chemical tricks and deceptions" in Wiegleb and Rosenthal's natural magic books.[18]

During the Renaissance invisible ink had become a curiosity without a cabinet, a wonder without a chamber. Although there might have been bottles of invisible ink hidden among the cabinets, they do not stand out and were never highlighted. While the Renaissance cabinets of curiosity were on the decline during the Enlightenment, invisible ink finally found its cabinet in these precursors to the chemistry set of the twentieth century. Like the curiosity cabinets of the Renaissance, the portable chemistry chests of the eighteenth century were objects of wonder and visually captivating. The natural magic tradition and these new portable chemistry cabinets would eventually lead to the design and marketing of the modern chemistry set.

By the beginning of the nineteenth century more and more chemistry cabinets became available in Germany and Britain, and they were pitched at an ever-broadening audience of ladies and gentlemen. The cabinet makers capitalized on the popular science movement, and their cabinets were used in well-attended public lectures and demonstrations. During the Enlightenment many of the popular science books focused on the physical and mathematical sciences and their illusions and effects, but by the early nineteenth century chemistry joined the trend in a big way. Jane Marcet's *Conversations in Chemistry* (1805) contributed to chemistry's popularity because it was written as a conversation and illustrated concepts using everyday examples.[19]

By the early nineteenth century it seemed as though "chemical recreations" had replaced the earlier more general rational and scientific recreations. In his wildly popular *Chemical Recreations* (which went through ten editions, beginning in 1834 and continuing through the nineteenth century), John J. Griffin included discussions of Hellot's sympathetic ink and some of the stories surrounding it. By the middle of the century the British company J. J. Griffin and Sons also started to manufacture and sell eleven different kinds of chemical cabinets, dominating the market for some fifty years until the outbreak of World War I.[20]

With the new natural magic craze, sympathetic ink had become a cornerstone of popular science. Wiegleb's publisher even wanted to impregnate several pages of the new natural magic book with sympathetic ink to help sell copies. Unfortunately, Wiegleb was too busy to procure the sometimes hard-to-find cobalt chloride, and Nicolai never followed up on the idea, even though Wiegleb recommended other chemists who might be able to find the cobalt (in our litigious age, such a suggestion would be vetoed because the reader might harm herself or the book might burn up).[21]

The early modern books of secrets and della Porta's "natural magic" had morphed into new natural magic books and "rational recreations" that began to resemble our modern notion of secular stage magic and parlor tricks. Unlike della Porta's natural magic, the magical recreations at the end of the century were more akin to the sleight of hand involved in pulling the rabbit out of a hat than to descriptions of the wonder of nature. They involved feats or parlor games that relied on the new science.

During the heyday of books of secrets in the sixteenth century, an invisible-ink recipe would occasionally be included, but it was written as a basic kitchen recipe with simple and available ingredients. For example, a recipe would suggest dipping an alum-written letter in water, where the writing would turn from white to black; or it might advise rubbing powder on fig-tree milk to make it visible. By the early Enlightenment period, Hellot had outlined the six classes of sympathetic ink, for each class specifying a way to make the invisible substance visible. The experimenter could paint, sponge, or spray a *liquid* on the secret letter, or wave it in the *air,* sprinkle *powder* on it, hold it over *fire,* or dip it into *water,* depending on what kind of ink was used to write the invisible message. Finally, there was the class Hellot made popular—the kind that appeared when heated and disappeared when cold. By the time the French populariz-ers and Wiegleb tackled the subject, the classes had been trans-formed into types with which one could develop neat parlor tricks or amusements. The changeable landscape was one of the first amusements to hit the streets of Paris, where it became very popular. Many others followed: the book of fate, the oracular letter, the magic portrait, the artificial hand, the wonder talis-man, the Sybil's or magic vase.

Much had happened in the development of invisible writ-ing since della Porta's day. The most innovative sympathetic inks had become part of natural magicians' repertoire. By the late seventeenth century one of the first "conjuror's ink"—the dangerous orpiment, "waters acting at a distance," the book-penetrating ink—had been quickly incorporated into ratio-nal recreations, with Hellot's sympathetic ink following on its heels. Both of these magical effects were featured in the newer "rational recreation" genre. Now, however, magic tricks featur-ing sympathetic ink began to be added, and they used illusion and science to entertain the audience or houseguests.

One popular sympathetic-ink magic trick was the so-called book of fate or fortune. Like a number of the other sympathetic-ink amusements, it involved writing a question in regular ink and an answer in sympathetic ink. The trickster was instructed to make a book of about seventy or eighty pages, with a secret compartment built into the back cover. The "magician" then wrote a question in regular ink, with an answer in sympathetic ink made from litharge of lead or bismuth. The visible questions were listed in the table of contents. The trickster soaked a double piece of paper in the so-called vivifying ink, made out of quicklime and orpiment, and placed it in the case at the end of the book. Then an audience member selected a question she wanted answered. The trickster placed the question on a piece of paper on top of the one written in the book, closed the book, pounded it shut, and placed a weight on it. When he opened the book, the "vivifying ink" had developed the answer. This feat illustrated sympathetic inks developed by liquids or the vapors from liquids.

Jean-Jacques Rousseau (1712–78), one of the most popular French philosophes, also experimented with the dangerous quicklime and orpiment solution. In 1736, when he was a young man and before Hellot published his findings on sympathetic ink, Rousseau tried to reproduce a sympathetic-ink experiment with directions from a professor of physics and probably with the help of Ozanam's *Mathematical Recreations.* After he had filled a bottle with the quicklime and orpiment mixture, it began to "effervesce violently." He ran to uncork the bottle, but it was too late. It burst in his face "like a bomb." He swallowed so much chalk and orpiment that it nearly killed him. He couldn't see for more than six weeks. His health also declined after this event. He felt "short of breath, had a feeling of oppression, sighed involuntarily, had palpitations of the heart, and spat blood; a slow fever supervened," from which he never recovered.[22] That is one reason we have not reproduced this experiment in the lab or in this book and would caution the reader not to try it out even though the early popular science books use the recipe freely and frequently.

Many of the recipes featured in the natural magic tradition slowly found their way into the growing world of chemical magic. Not surprisingly, the new chemical magic phenomenon coincided with the height of Victorian parlor magic in the middle of the nineteenth century and was an outgrowth of the natural magic and rational recreations movement. Unlike the other sciences, however, chemistry became uniquely suited to performing magic. Magicians encouraged conjurors to incorporate chemistry experiments into their parlor magic. Magicians' manuals like *The Magicians' Own Book* encouraged amateur magicians to learn about optical, mechanical, or "magnetical" illusions. By the end of the nineteenth century, chemical magic had developed into a special scientific kind of magic in Germany, America, and Britain.

In the early twentieth century many magicians had become "scientific conjurors," and "Professor" Ellis Stanyon offered courses on "Fire and Chemical Magic" at his "School of Magic" in West Hampstead. Unlike the Hogwarts School of Witchcraft and Wizardry of Harry Potter fame, Stanyon's school of magic emphasized scientific wizardry and operated without witchcraft. Even so, Stanyon thought that they had achieved "real magic at last" with dramatic color changes.[23]

In 1909, the magician William Linnett presented a paper at the Society of American Magicians declaring that chemistry was uniquely adaptable to magic and recommended that all magicians become familiar with chemistry.[24] Many magicians took his advice to heart, while many chemists became enchanted with the magical color changes. John D. Lippy, a chemist and amateur magician, wrote a manual for magicians called *Chemical Magic* and dedicated it to "all those magicians who have made magic an art." By the time of the American edition (published in 1959), teachers were recommending the book as a "very fine reference for ambitious teachers who want to popularize scientific information by presenting dramatic experiments."[25]

Lippy also asked the "world famous magician" Harry Blackstone to write some prefatory remarks for the American edition. In that introduction Blackstone opens with the alchemists' quest just as the magic book accompanying the Chemcraft chemistry set had many years before, but he notes that the world is now built on "physical principles," not the "black arts." Still, he argues, "magic stays with us—not black magic, but a magic that entertains . . . magic that, at its best, thrills"; the magician "can unlock the secrets of chemistry." Even though Blackstone rejects the alchemist and his rigmarole, he still thinks that alchemy has value because its "two heirs" are chemistry and magic.[26]

When the Chemcraft chemistry set featured the Oriental alchemist, it was incorporating a magical tradition while adding mystery. The orientalist theme was common in professional stage magic at the turn of the twentieth century. Even so, Chemcraft's suggestion to use a made-up Ethiopian slave, "his face and arms . . . blackened with burnt cork," and bearing a "fantastic" name like Allah, Kola, or Rota as an assistant, sounds offensive to twenty-first-century ears.[27]

In addition to drawing on a magical tradition, Chemcraft chemistry sets emerged out of a new political context at the beginning of the twentieth century. J. J. Griffin and Sons had dominated the chemical cabinet market until World War I, at which point Germany stopped exporting the necessary chemicals and England and France redirected their limited chemical resources. In 1914, as hostilities began in Europe, the American Porter Chemical Company stepped in and transformed the Victorian chemistry cabinets into a toy—the twentieth-century chemistry set. Harold Porter and John J. Porter set up the Chemcraft line, which resembled the chemical magic tradition from the Victorian period. The earlier sets were made of wood, but they also shared the side doors so characteristic of the later metal chemistry sets. Even though the chemistry set started with a magical tradition, it soon developed the needed niche of an educational

Chemistry set, Kristi Eide, photograph

toy.[28] And it was on this magical toy that invisible ink piggy-backed into twentieth-century America and enchanted scores of young children with its magical color changes.

"The Gold-Bug"

The natural magic tradition, Victorian scientific and chemical recreations, and secular stage magic had all helped to propel invisible ink into the world of magic during the nineteenth century, but magical ink wasn't confined to magic and science; it was beginning to make a lasting impression on literature and culture.

By the middle of the nineteenth century many Americans had also become fascinated with codes, ciphers, and invisible ink. The subject peppered the pages of popular magazines like *Graham's* and *Dollar*. Edgar Allan Poe (1809–49) was one of the most prominent Americans intrigued by ciphers and invisible ink. Although Poe is remembered for his gothic, brooding poems and short stories, he also wrote several articles about secret writing. He was gratified by the widespread interest in the subject and sponsored contests challenging his readers to send him codes to break. He also capitalized on this interest in his short story "The Gold-Bug." More importantly, he stimulated interest in the subject among prominent cryptographers decades later. William F. Friedman, the great American cryptographer who broke the Japanese purple code during World War II, credits Poe's short story with inspiring his interest in secret writing. When he was a child, he "would talk with excitement about a world he had discovered through Edgar Allan Poe's 'The Gold-Bug.'" He was so enchanted by Poe's story of buried treasure that for the rest of his life he was still "prepared to waste time" on any such "messages sent to him for decipherment."[29] According to the eminent Poe scholar Thomas Ollive Mabbott, "The Gold-Bug" was one of the most popular and imitated short stories in history, inspiring hundreds of spinoffs and films. For example, Robert Louis Stevenson's book *Treasure Island*, published in 1883, was clearly inspired by the theme of hunting for buried treasure. There was also allegedly even a true nineteenth-century story about the hunt for a lost treasure described in the 1885 pamphlet entitled *The Beale Papers* about a treasure buried in 1819 and 1821 in Virginia that has never been recovered. Stimulated by the Beale Cipher, treasure hunters continue to look for the lost treasure.[30] Of course, these earlier stories could have inspired both Poe and Stevenson.

"The Gold-Bug," first published in *Dollar* magazine in 1843, is a story about recluse William Legrand's quest for a hidden treasure buried near Sullivan's Island, South Carolina, where he lives in a hut with his black slave, Jupiter. When Legrand finds a striking gold beetle and a piece of parchment marked with a skull's head on a deserted beach on the mainland near his hut, he hopes the gold beetle itself will lead him out of his newfound poverty. By the time the narrator of the story visits Legrand, he has lent the beetle to a naturalist on the island and stuffed the parchment in his pocket. Excited about his find, Legrand wants to show an image of the beetle to his friend and visitor; the unnamed narrator draws it on the parchment. When the narrator examines the parchment, he holds it up to the fire to look at the drawing more closely, but instead of a beetle he sees a skull or death's head. Unbeknownst to Legrand and the narrator, the skull had been drawn in invisible ink on the opposite side of the paper, to be revealed only when the narrator placed the parchment in front of the fire in order to see it more clearly.

Legrand begins to suspect that heat has brought the image to light. Later he tells the narrator, "You are well aware that chemical preparations exist, and have existed since time out of mind, by means of which it is possible to write . . . so that the characters shall become visible when subjected to the action of fire." Legrand is familiar, as was Poe, with the effects of zaffre and the regulus of cobalt. According to Legrand, when the former is dissolved in hydrochloric acid and diluted with water, it produces a green tint, while the latter produces a red color when dissolved in nitric acid. Author and character are also both aware that the cobalt solutions produce inks that appear when subjected to heat and disappear when cooled. Soon Legrand realizes that there might be more written on the paper, and he "immediately kindle[s] the fire, and subject[s] every portion of the parchment to glowing heat."[31]

The first visible signs are a figure of a kid, but there is still a big gap between the skull stamp and Kidd's signature, and Legrand suspects the secret memorandum is written between the symbols. Legrand concludes that the treasure seekers lost the instructions for finding the buried treasure, and that this thin piece of parchment may have hidden instructions on it. He persists. He holds the vellum in front of the fire, but nothing appears. He then rinses the parchment with warm water and places it over a hot tin with the skull facing downward. To Legrand's "inexpressible joy," red characters start to appear in straight lines, producing a ciphered text.[32]

When he finally deciphers the four-line text, it specifies the location of the buried treasure. The magical quality of the adventure is heightened by Poe's description of Legrand as he walks to the treasure site, carrying the beetle by attaching it to the end of the whipcord, "twirling it to and fro, with the air of a conjuror as he went."[33]

Poe's story is usually referred to in accounts of ciphers in literature; the invisible ink is barely mentioned. Legrand solves the puzzle of the odd skull by using the skills of a detective as he re-creates all the events that led to its appearance and concludes that the paper had been warmed by heat when the narrator examined it. This clue leads him to discover the existence of an invisible-ink message. In other words, the secret ciphered message would never have been discovered if Legrand had not heated the paper in order to reveal the hidden message. More than simply a fictional story about ciphers, "The Gold-Bug" illustrates very nicely how ciphers and invisible ink work hand in hand to create the best possible, and hardest to break, encryption.

Hellot's discovery of cobalt sympathetic ink in 1737 marked the beginning of a new period in the development of invisible ink. Until that time, prisoners, lovers, and spies had used the substance primarily to conceal secret messages. They needed invisible ink to hide secret plans of escape, to plot against the government, or to pass on secret love letters. The secret ink they used was fairly primitive. The discovery of a magical ink that both appeared and disappeared captured the imagination of a new set of people and stimulated new applications during the eighteenth century. It intrigued scientists, inspired poets, and stimulated parlor tricks. During the hundred years separating the lives of Jean Hellot and Edgar Allan Poe, there was an enormous growth of public interest in science and in secret writing. Cobalt sympathetic ink had migrated from the elite pages of the French Royal Academy of Sciences to the popular press of the *Dollar* magazine.

Around Poe's time and well into the late nineteenth century, cobalt sympathetic ink also became a part of popular and serious chemistry texts. Chemists used the color changes and the changeable landscapes to illustrate chemical principles while entertaining the reader. By this time they knew that cobalt chloride lost a water molecule when heated (an anhydrous state) in order to turn the beautiful blue color.

Even though secret ink had gone public, no innovative sympathetic-ink classes emerged, and no one seemed to have developed new chemical combinations during the nineteenth century. Instead, it became the cobalt century. Everywhere people turned, they continued to be enchanted by the majesty of cobalt, by the changeable landscapes, by the weather dolls, by the magic tricks, and by the simple beauty of the magical color changes that appeared before their eyes like a spreading turquoise Caribbean ocean on white sand.

It has been my contention that the serious use and development of secret writing increased and intensified during periods of war. Invisible writing was considered an essential part of secret communication. However, in addition to its association with international intrigue, invisible ink conjures up links to worlds of wonder and magic. And this feature of invisible ink is no less important to humanity. For scientists, wonder and magic are often the beginning of a passion for the natural world. Many scientists have claimed that it was the childhood chemistry set, including the magic of invisible ink, that stimulated their interest in science. They weren't alone. The magic of cobalt sympathetic ink had captivated visual artists and provided magical playthings for the home, like the changeable fire screens and weather dolls, but it also inspired the literary and poetic imagination and entered the world of popular science and stage magic.

The Secret-Ink War

Scene I

By the time Carl Frederick Muller faced the firing squad at the Tower of London, he was calm. It was 6:00 AM on June 23, 1915, and he shook hands individually with all eight members of the firing party at the Miniature Rifle Range, telling them that he understood they were performing their duty. The surgeon-colonel led him to the chair of death, which was tied to stakes driven into the ground. Muller sat on it quietly as the surgeon-colonel buckled the leather strap around his waist and blindfolded him. The firing squad raised their rifles and shot Muller in the chest. He died instantly from shock as the bullets passed through his thorax and out of his body.[1]

But the night before his execution, the fifty-eight-year-old Baltic German whose German name was Karl Müller was a nervous wreck. He sobbed through the night, asking for his wife and children.[2] By the time the British had arrested the tall man in a frock coat, he seemed to have a perpetually worried look on his long, gaunt face. His walrus mustache made him look even sadder. Muller denied that he was a spy for the German secret service, but his background made him the perfect spy.

Carl Muller

Muller wasn't the first German spy to be shot during World War I, nor was he the last. One after another, in rapid succession, during the years 1914–16 the Scots Guardsmen shot eleven accused German spies (nine of them in 1915 alone). It's hard to believe, but Muller's fate, along with the fates of several other German spies, hung on the use of a lemon.

Scene II

There were so many bags of mail piled up in the office of British Postal Censorship during World War I that examiners waded through and tripped over them in order to grab a promising one for examination. Then the contents of the bags turned into stacks of papers piled up on examiners' desks. By the end of the war, millions of pieces of mail had been examined every month by some forty-two hundred men and women. When war broke out in 1914, only seven people had worked in Censorship. In addition to its job of traditional censorship—removing or obliterating military secrets—the British began to use mail interception for counterespionage purposes by looking for secret messages in the mail.

About three thousand of these censors were women. Part of the reason for the large number of women engaged in censorship was that so many men had been drafted into the army. Women also served as a cheap labor force, as they earned considerably less than their male counterparts. But many British participants in the battle against foreign spies also thought women had a sixth sense for detecting suspicious letters. The women quickly learned that a letter that said absolutely nothing should be considered suspicious. And nine out of ten times they were right.[3]

Although so-called Black Chambers had existed in Europe since the 1600s and examiners had filtered a hundred letters a day, British Postal Censorship during World War I was probably the largest and most successful effort in history to monitor and intercept mail and cables.[4] It was truly imperial, as its tentacles reached from England and Europe to the horn of Africa and the Caribbean.

The key to the examiners' success in detecting spies was monitoring cover addresses in neutral countries. Otherwise, it would truly have been a needle-in-a-haystack job, since 375,000 pieces of mail were hauled in every day—millions of pieces a month. When MI5, Britain's new military counterintelligence service, learned about a new German cover address, it was placed on a black list, and everyone who wrote to this address was watched. By the end of the war there were 13,524 names on this list.[5]

The Lemon Juice Spies

Like many other German spies, Carl Muller became a victim of the counterespionage efforts of British Censorship. Born to German parents in 1857 in Russia's Baltic port city Libau, Carl Muller claimed that he lost his parents at a young age and was brought up by his uncle, who was mayor of the city. In addition to the Russian and German he learned in Libau, Muller also acquired Dutch, Flemish, and English through his travels with shipping companies.

Muller went to sea at the age of sixteen, and his seafaring career took him from a job at the American Shipping Company in Hamburg to a stint in New York City with his new Norwegian wife. He finally moved to Antwerp, Belgium, with his family and opened a boardinghouse. After his wife died he ended up in Rotterdam, Holland, working for an English company checking cargos. After a mysterious illness, he returned to Antwerp, where he went into business with an Englishman as a cargo superintendent dealing with German steamers. Everyone thought he was German.[6]

Germany invaded Belgium on August 4, 1914, and troops quickly occupied the neutral country. By October, when German forces bombarded Antwerp, only about five thousand people still lived there. German soldiers were looking for lodging and supplies and a place to live. Soon many of them ended up at Muller's house.[7]

It is likely that the German Secret Service recruited Muller during this time, because occupied Antwerp had become a center of German spy activity. It was home of the War Intelligence Center (*Kriegsnachrichtenstelle*), the major base for operations against Britain, and the famous spy school headed by the legendary Dr. Elsbeth Schragmüller, more popularly known as the Fräulein Doktor, a mysterious blonde woman with a Ph.D.

Muller himself never revealed any information about his recruitment and denied he was a spy until the very day he was shot by the firing squad.

On January 9, 1915, Muller boarded a ship in Antwerp and sailed across the English Channel to Hull, England. Before he settled into a boardinghouse in London, he looked up old contacts and friends who could help him establish a spy network.

Within three weeks Muller was communicating secretly with his handlers. On February 3 he wrote a cover letter in black ink to a fictitious client, ordering that the delivery of goods be postponed. But between the lines of the innocent letter, he reported on British troop movements—fifteen thousand English troops had apparently left for France via Southampton. Muller didn't know that British Postal Censorship intercepted the letter.[8] The next day he wrote another secret-ink letter. It was also intercepted. The same day he left for another trip to Rotterdam. Since Rotterdam was without a doubt "the spy capital of the Netherlands," it is likely he met with his handlers there after the initial training in Antwerp. With "its port, regular shipping to England and a large German community," and its neutrality, Rotterdam was the ideal place for exchanging documents or meeting with contacts.[9]

A few days after Muller's return to the 38 Guildford Street boardinghouse where he was staying in February, Inspector Buckley from the Criminal Investigation Department at New Scotland Yard showed up with a search warrant. Although he was not acting on the intercepted letters, the inspector had heard that Muller was a spy. But the search turned up no evidence and the case was filed away.

Surprisingly, the Scotland Yard visit didn't scare Muller into curtailing his spy activity. A few days after the inspector's visit, he wrote yet another covert letter. In the meantime, he also contacted a baker in London, John Hahn, whom he had met in Antwerp, as a step toward expanding his spy network. This would prove to be a grave mistake.

On Muller's first visit to the bakery, Hahn, a short, heavy, clean-shaven man, met him at the front of the shop, but they exchanged only a few words because Hahn was working with his dough and couldn't talk. Regretting that he had rebuffed Muller, Hahn later invited him back. Muller then showed up unexpectedly for tea the following Sunday.[10]

Hahn seemed like an ideal candidate to work as a spy for Muller. He was born and raised in England, but his parents were German. He learned the trade of baker during a stint in Breslau, Germany. When he returned to Britain he had opened a bakery but then fell on hard times and had to borrow money from his wife's father. When war broke out, there was considerable anti-German sentiment and Hahn's shop was attacked during the riots. Short of money and bitter about his experiences, Muller's advances were accepted. But there was one problem Muller hadn't anticipated.

By the end of February, Muller was teaching Hahn how to "write in lemon." Muller told Hahn to write a letter to a contact on the Continent who might be able to help him obtain a position. Hahn wrote this in plain text while Muller wrote a secret letter in German with lemon juice. When his wife left the office, Hahn got a lemon from the bakery and brought it back upstairs. Muller "showed him how to put it on the fire."[11] Hahn then mailed the letter at the mailbox near his house. It was quickly intercepted.

While Muller contacted a friend in Sunderland, Hahn decided to write a letter to the job contact on his own. He also wrote a secret-ink letter on the back with lemon juice detailing that four thousand to five thousand men were stationed at the Manchester Canal. He referred to Muller's code name "A.E. 111" and signed the secret letter with his real name! When the British post office intercepted the letter and found secret writing, it was forwarded to Sir Basil Thomson at New Scotland Yard. It was easy to find John Hahn.

On the day they read the letter, New Scotland Yard sent inspector George Riley of the special branch to search the bakery and talk to Hahn. Because her husband was out, Mrs. Hahn let Riley in to examine the premises. He located a chest of drawers with a green penholder on the desk and a piece of lemon in an inside drawer. It was still juicy and pierced at the top. Riley also found blotting paper and a nib hidden behind a clock. When he held the blotting paper up to the light, he found indentations with Muller's street address.

Now Riley had to wait until 10:30 PM before Hahn came home. When Hahn arrived, Riley asked him about his associates. As Hahn told him about Muller, Riley dramatically pulled out the lemon. "The lemon!" Hahn exclaimed, quite surprised. He was arrested and taken to the police station for questioning.

Though Hahn had seemed a needy spy candidate, Muller failed to notice one thing: he was a simple and honest man, not cut out to work in the shadows.

The next day another inspector, Edward Parker, was sent to Muller's residence on Guildford Street. After a half-hour wait, Muller arrived, and the inspector arrested him on suspicion of espionage.

Parker and the police officers searched Muller's room and found correspondence and a memo book with addresses. Most tellingly they also found three pieces of lemon and cotton wool in a dresser drawer, together with two pens. Muller protested and demanded to know the meaning of the search. He extended his shaky hands and described his apparently nerve-racking ordeal in Antwerp. Instead of offering sympathy, the police ser-

C. A. Mitchell

geant searched and found a lemon in Muller's overcoat pocket. When Parker asked Muller what the lemon was for, he pointed to his mouth and said: "My teeth."[12]

Needless to say, using lemons to clean teeth instead of corroding them was as laughingly unconvincing in 1915 as it would be in the twenty-first century! In order to examine the lemon juice evidence, the Censorship office brought in Charles Ainsworth Mitchell (1867–1948), an expert on the chemistry of inks, who would later become one of England's most respected forensic scientists. With his friendly personality, he became a "pillar" of the new Society of Analysts. Though he spent most of his time squinting through microscopes and writing numerous books on science and the criminal, he had time to testify in court on cases ranging from murder to poisoning, wills, forgeries, and the World War I lemon juice spies. A kind man with an artistic temperament, he even wrote some short stories and poems.[13]

Mitchell spent the next few days examining the evidence retrieved from the Muller and Hahn homes. He tested both the

Black lemon

pens and the lemons, testing for the presence of iron. When he examined the Manos pen from Muller's room, he found that it was corroded from the acid in the lemon juice and that it contained iron. It also had cellular matter from the lemon on it. When he tested pieces of the lemon, he also got reactions suggesting that the spy had pushed the nib in and out of the lemon to write with it. When Mitchell examined Hahn's nib he found that it had been lacquered to prevent corrosion, but it, too, had cellular matter from the lemon on it.[14] The fresh, juicy lemon was prime evidence in Muller's and Hahn's trial. This lemon still exists in the British National Archives. Almost one hundred years later, it is no longer yellow and juicy, but rather black and shriveled.

After their arrests, Muller and Hahn were taken individually to the Tower of London, each unaware of the other's arrest. In addition to the exhibit evidence, like the intercepted secret spy letters, the lemon juice used to write them, and the juicy yellow lemon, British authorities took eleven days' worth of evidence from a parade of witnesses who had had contact with the spies.[15]

Muller pleaded not guilty to the charge of espionage, while Hahn pleaded guilty. Muller stated that he had merely been collecting information for a journalist. He claimed that the information had all been trash and imagination and that the evidence that would vindicate him was in Antwerp. Authorities did not believe a word of this. Muller was considered a dangerous spy. He was sentenced in the Central Criminal Court to be shot. Hahn was given a much lighter sentence of penal servitude for seven years because of the circumstances surrounding his recruitment.[16]

On June 22, 1915, Muller was loaded into a taxi to be taken from Brixton Prison to the Tower of London for his execution. But the cab broke down during the lunch hour and attracted a crowd. Seeing a foreigner sandwiched between two uniformed policemen driving toward the Tower, the crowd started shouting "German spy!" But another taxi was quickly found, and Muller was taken to the Tower of London's Miniature Rifle Range without further incident.[17]

Since Muller's trial was held secretly, the German Secret Service didn't hear about his execution for a long time even though there was a passing reference to Muller and Hahn in the widely publicized Anton Küpferle case. The Germans continued to send payment for his services, and the British continued to send false information under his name. With the money realized from the German Secret Service, the British officers were quite proud of buying a car they dubbed "The Muller."[18]

British investigators convicted 30 spies out of about 120 who seemed to be active in Britain from 1914 to 1918. Muller was the second spy to be shot in the Tower of London. The first spy to be shot in the Tower in 150 years had been the infamous Carl Lody.

Unlike Muller's secret trial and execution, Lody's case made headline news. His public trial and execution were certainly ways to set an example and to supply a deterrent to prospective Germans spies. But there was more. British authorities were charmed by Lody. Although he pleaded not guilty to the charge of espionage, he presented himself as a patriotic "man of honor," not a lowly spy. He wrote to his sister Hanna that he was dying "in the service of the fatherland," which rendered death easier for him. "I will die as an officer, not as a spy."[19]

When his time had come, Lody asked the assistant provost-marshal: "I suppose you will not care to shake hands with a German spy?" "No, I would not," said the A.P.M., "but I will shake hands with a brave man," and so Carl Lody died before the firing squad at 7 AM on Friday, November 6, 1914.[20]

As with Muller, British authorities discovered Lody because of mail interception by Postal Censorship. His mistake was that he wrote to one of the German cover addresses that the British had under surveillance. The Germans had numerous cover addresses of innocent people they used to pass on mail. The British had quickly developed a black list of these German cover addresses.[21]

It might have been because Lody apparently did not use secret writing (at least it wasn't showcased as evidence) that nearly half a dozen spies after him were told to use invisible ink. But their spymasters did not provide them with sophisticated invisible-ink formulas, because a number of lemon juice spies were caught in 1915. Investigators found more sophisticated invisible ink in only one case during the early years of the war.

Two of the most striking German spies who used invisible ink did not end up on death row. But this was not because the British felt kinder toward them or found their deeds less nefarious.

Anton Küpferle claimed to be a naturalized American citizen from Switzerland who changed his name to Anthony Copperlee in 1912. In fact, he was born in Germany, and his father may have brought him to the United States of America when he was nine years old. He seems to have gone back and forth between the continents before getting himself ready to leave for England in February 1915. He had acquired an American passport in January and allegedly left behind a business as a wool merchant. The British did not like him and called him a "stupid Prussian" with "stiff, upstanding hair, round spectacles," and a "painfully assumed American accent."[22] It didn't help that he also endorsed the use of poison gas warfare against British soldiers because of the "stupefying death" it caused.

Küpferle was the first agent to be caught through the German intelligence Rotterdam cover address under surveillance by the British. Examiners ran a heated flatiron over the three intercepted letters and found messages written in secret ink that contained "information of military and naval importance." The British had an easy time finding Küpferle because he used his real name in the letters and included his address at the Wilton Hotel, London.[23]

From the time he arrived in Liverpool on February 14, British counterintelligence watched his movements and mail and gathered evidence to make a case against him. The British suspected him because they had intercepted a seemingly harmless letter he had written to the same cover address that Muller and Hahn had used. One of those especially talented, eagle-eyed woman censors with a sixth sense apparently flagged a letter. Sure enough, between the lines of a letter that announced he had arrived safely in Liverpool, planned to travel to London the next day, and expected to arrive in Rotterdam later that week, was a chilling description of war vessels he had seen while crossing the Atlantic.[24]

The woman was Mabel Beatrice Elliott, who had applied heat

Mabel Elliott

to the letter and unmasked the spy Küpferle. She also discovered secret messages in Muller's and Hahn's letters. When she made these discoveries she was a deputy censor examiner, but her skill, keen observation, and organizational abilities were so valued that she was promoted to assistant censor of Postal Censorship. The next year she was chosen to oversee all three thousand women in the program. Even though Elliott was honored with an M.B.E. and other awards, her role in breaking this spy ring was never mentioned during her lifetime.[25]

Within days, police arrested Küpferle and informed him that he was suspected of espionage. When they took him to Scotland Yard, investigators found the same paper he had used in Liverpool. He also had two lemons in his bag, along with a bottle of formalin labeled "L. Friesch, Deutsche Apothecary, Brooklyn, NY." To top off the evidence, they found a pen in his vest pocket with lemon juice still stuck on the nib. Expert lemon juice spy catcher Charles Ainsworth Mitchell was brought in to examine the evidence.

After the first day of the trial—Elliott provided evidence under the assumed name Maud Phillips—Küpferle felt doomed. He would not sit through a second day. Even though he was under a twenty-four-hour watch, the night warden found him hanging from the wall grate. He had wrapped a silk handkerchief around his neck and tied it to the metal grate, stood on a book from the library, then kicked it away. His body was still warm, but guards could not revive him.

He left behind an astonishing confession on his prison slate. He didn't want to lie anymore and wrote that he was born in Söllingen, Rastatt, Baden, and that he was a soldier in the German army. Like Lody, he declared that he was "not dying as a spy, but as a soldier."[26]

Küpferle's case and the dramatic prison suicide made headline news. Now the whole world knew about his suicide, the prison slate suicide note, and his use of lemon juice and formalin.[27]

At least four of the eleven German spies shot in the Tower of London used lemon juice to pass secret messages. A lemon or lemon juice was often prime evidence in convicting them. It is, of course, surprising that the scientifically advanced Germans used such a primitive form of invisible ink. Not only was lemon juice as invisible ink known to most schoolchildren, but it had been the standby used for some three hundred years. All one needed was the heat from a candle or iron to reveal the image, even with formalin added, which made it harder to reveal.

Spies during World War I also wrote secret messages with other organic sympathetic inks, like vinegar, onion juice, saliva, urine, milk, semen, and other readily available household substances.

We can only speculate why the Germans did not provide these early spies with more sophisticated writing methods. Probably the Germans were unaware early in the war how sophisticated and all-pervasive British Censorship had become. After all, un-less a letter was suspect, no visible spy message was seen.

After the painful and visible loss of these 1915 lemon juice spies, the Germans began to develop more sophisticated invis-ible-ink methods. But before that adjustment, a music hall per-former, Courtney de Rysbach, confused investigators about the type of ink he used to write between the lines of music.

De Rysbach, a British citizen of Austrian descent, was a well-known vaudeville artist in England. He was a bit of a comedian, he juggled and sang a little, and he could do tricks on a bicy-cle. He was touring Germany as a dancer and performer at the outbreak of war when he was detained at the Berlin-Ruhleben prison camp, along with other foreigners. While he was there, the German Secret Service offered him freedom in exchange for spying for them in Britain. He accepted, but he later claimed he intended to collect only "trash."[28]

In July 1915, the British Postal Censorship department inter-cepted four sheets of music with clear signs of invisible ink between the bars of music. One of these was the musical score "The Lad-der of Love," and the other "On the Way to Dublin Town." De Rysbach constantly asked for money, and on the "Ladder of Love" score he wrote invisibly that he needed money to obtain naval se-crets from his brother. On another piece of music he complained that he had asked for money in "24 newspapers and 8 letters and 2 telegrams," to no avail. He apparently started writing between the bars of music after discovering, as he wrote to his handlers, that "the newspaper trick is found out and very carefully watched."[29]

At first investigators thought de Rysbach, like the other Ger-man spies, had used lemon juice as an invisible ink because they found half a lemon on the hotel room mantelpiece of his fiancée, Miss Ena Graham. The other half was found in the bathroom and had been used to make lemonade. Investigators also found a nib in Miss Graham's suitcase.

Secret ink on music

The lemons were clearly a mystery. When Curtis Bennett cross-examined the manager of Hotel Sutties, he asked her: "Tell me will you, I do not want to have any mystery about these lemons which have been mentioned. You know, do you, that those lemons were actually squeezed into a glass in the dining room and used for lemonade?"

The manager testified that she squeezed the lemons herself, and the waitress had taken the lemonade into the dining room, where de Rysbach and Miss Graham had drunk them. The prosecutors still doubted that the waitresses remembered correctly. They assumed that de Rysbach somehow had secreted the juice to write lemon letters.[30]

When Mitchell was brought in to testify, he claimed that there was lemon cellular matter and acid on the nibs, though he was not asked to examine the musical scores themselves. He concluded that the nibs had been used "for writing with lemon juice." De Rysbach simply stated: "All I want to say is, he is wrong."

Later de Rysbach admitted that he had written invisibly with pomade, a French brand called Oja, which investigators found in his suitcase. Pomade is a greasy dressing that leaves hair slick and shiny. People usually associate the pomade-created hairstyle with 1950s rock stars like Elvis Presley. Judging from pictures of de Rysbach, he seems to have used the substance on his hair as well as sending secret messages with it.

Pomade, a petroleum-based substance, does not work by itself as invisible ink. But the German Secret Service had begun to disguise its invisible ink in household items like medicine, soap, toothpaste, mouthwash, perfume, and even pomade. The British never found out what was mixed in with the pomade, but regular post office chemists developed the secret writing by smearing the invisible ink with a combination of black ink and sulfide of iron.

The censor's chemists were never positive whether the invisible ink on the sheet music had pomade on it or was simply lemon juice. When John Price Millington, a chemist with a degree from Cambridge University, was questioned by several of the prosecutors, he really wasn't sure: "It would be very difficult to say positively what was used because many materials which can be used show up . . . when treated in this fashion . . . : like 'fruit juice.'"

When the examiner pushed Millington and asked whether he could tell what the substance was, he waffled: "No, I do not think—yes I think I could my Lord. I think I should say that that had not been written with lemon juice, except under extraordinary circumstances."[31]

It turned out that when the pomade was heated, the yellowish brown color wasn't as dark as the charred brown color lemon juice creates. But when it was smeared with the ink and iron sulfide solution, it was much easier to see.

To complicate matters, British Censorship's testing department concluded that de Rysbach's secret ink had been prepared from toothpaste impregnated with potassium ferrocyanide. Whatever the substance really was, the case marks a transition between the simple lemon juice spies to more sophisticated cases that were to challenge British chemists.

Even though de Rysbach was found guilty, he wasn't executed as his predecessors had been, in part because there was not enough evidence to contradict his claim that the Germans had interned him. He was sentenced to life imprisonment.

Censorship Grows

There is no doubt that Postal Censorship—especially its mail interception component—played a huge role in catching German spies. It became a major part of Britain's counterespionage activity, along with the efforts of the New Scotland Yard and the

Secret-ink research laboratory

newly formed MI5 during the early years of the war. MI5 and the New Scotland Yard investigated Muller, Hahn, and Küpferle because an informant had told MI5 about a man living in Rotterdam who was a German agent. His mailing address was a box used by the German consulate for intelligence purposes. MI5 intercepted and examined all mail sent to this address and arrested all three men within a month.[32]

After these early successes, Postal Censorship moved into the War Office and quickly set up chemical and code branches—usually called the testing department—to support censors in their work examining suspicious mail. As early as the end of November 1914 an experimental scientist began work as a trade

Secret-ink testing room

assistant censor. Chemists and physicists were hired, and an elaborate system of paperwork and slips of paper was established to organize the knee-high bags and bags of mail diverted to censors.[33]

British Postal Censorship was truly an imperial enterprise. We can read of British imperialism, we can learn about Britain's colonies, but it is hard to fathom the extent of Britain's Censorship empire without looking at correspondence documenting its reach. A dizzying amount of letterhead stamped "Imperial Censorship" litters the archives from Britain's colonies in the Caribbean and Bermuda, in the Horn of Africa, in Australia, and in the Middle East.

Testers at work

The British themselves were well aware of the influence and power of the office. Sidney Theodore Felstead, a British journalist and author, made this comment in 1920: "The great importance of the postal censorship . . . lay in the grip it gave us over practically the whole of the world's correspondence."[34] Here was a worldwide surveillance program predating by almost a century the advent of the World Wide Web, the electronic echelon system, and the National Security Agency's global reach! Remarkably, the British conducted their worldwide surveillance manually.

Dr. Stanley W. Collins (1882–1954) became legendary as a secret-ink detection expert when he was head of Postal Censorship's chemical department. The London-based Collins was educated at King's College, London, and served the Postal Censorship department in the War Office as chief scientific officer in both world wars. His secret work was so valued that he was given the Order of the British Empire (OBE), decorated with the Palmes d'Officier d'Académie in 1920, and awarded the liberty cross by Norway in 1947. Despite his renown within secret-ink circles, his life remained largely invisible. No published pictures of the man can be found. No major newspaper published an obituary of him, and he published no major book or article about his secret war work, though he was writing about the testing department with "fervour and enjoyment" before he died in 1954.[35] His study never saw the light of day; maybe he wrote it in secret ink . . .

Collins owes his public reputation as England's leading secret-ink expert, in part, to the American code breaker Herbert Yardley's best-selling tell-all book *The American Black Chamber,* first published in 1931. Yardley and his fledgling staff in the U.S. War Department found Collins to be a friendly man who entertained them with secret-ink spy stories when he visited Washington, D.C., for two months in the summer of 1918, several months after America had entered World War I. The War Department had brought Collins to help Yardley and the American codes and ciphers section set up a new secret-ink lab. Until 1931 no one had heard of Collins. Unlike the public analyst C. A. Mitchell, he wasn't in the limelight because of his publications. But Collins's lectures on secret ink were well illustrated with cases of German spies in England.[36]

In fact, Collins was sent to the United States as an instructor to U.S. military intelligence after several American military intelligence officers had visited British laboratories in 1917. Collins first landed in New York in early May 1918 and visited the Americans' equivalent of Postal Censorship, in Lower Manhattan, run by Captain Emmett K. Carver. The lab there was about as big as a bathroom.

At the end of May, Collins proceeded to military intelligence headquarters in Washington, D.C., where he met with its chief, Colonel Ralph Van Deman, who was reorganizing the War Department's military intelligence division and was eager to create a scientific department. Collins gave his lively lectures in front of about 150 officers there.

Collins then returned to New York and expanded Carver's bathroom lab with a fifty-foot-by-thirty-foot testing room and a darkroom, leaving the little room for special work. The department added a chemist and five women assistants. Another separate but similar lab setup was also created in Washington, D.C.

When Collins finished his advisory tour in the United States in early August and returned to England, Captain Carver tagged along to obtain further information and advice on setting up an effective testing department.[37]

Even though Collins shared a lot of his secret-ink knowledge with Yardley, he never mentioned the German lemon juice spies. He may not have known a lot about the specific cases because he had been brought in soon after the Germans began to improve their secret-ink methods. Not only did the Germans begin to devise more sophisticated invisible-ink formulas than those based on fruit juice, but they also began to conceal the liquid in medicine bottles, soap, and other household articles. And this is how the secret-ink "seesaw war" began.

As soon as the British found a way to develop the secret messages, the Germans countered with better concealments and more secure secret formulas. At first the British "striped" suspicious letters with three or four brushes impregnated with three or four different reagents—sometimes called "Britannia's Trident."[38] If the letter had secret writing on it, one of the developers usually worked. But as soon as the British started using this method, the Germans countered by devising secrets inks that needed *one* specific developer, or reagent, to reveal their secrets. In other words, most insecure inks can be developed by a variety of chemicals; the trick is to find a chemical that can be developed by very few chemicals or, even better, only one chemical. This presented a great challenge to the Allied secret-ink sorcerers, who dreamed of creating a *universal* reagent that would develop all secret writing.

The first step toward creating a universal method was the iodine vapor test. The French, British, and Germans all claimed to have developed this method during World War I, but in fact Belgian legal chemists discovered it in the late nineteenth century in their efforts to uncover fraud committed on writing paper. Gustave Bruylants (1850–1925), a noted Belgian chemist who had studied with the German organic chemist August Kekulé, one of the most prominent chemists in Europe at the time, and his colleague Léon Gody, a professor at the Military College, noticed that when they applied iodine vapors to paper, it turned blue-violet when it landed on the disturbed parts of the paper.[39] (See the kitchen chemistry appendix.)

The iodine vapor test worked by depositing iodine on any part of the paper written on or scratched by a secret-ink writer. It was also used to detect fingerprints on rough surfaces. Anytime someone writes on paper, it disturbs the fibers. Using this method, the British were able to read many secret-ink letters. But all of a sudden, the method stopped working. The Germans had countered by telling agents to smooth the paper fibers with steam or an iron before they sent the message, to dip it in water, or else to write a message on another piece of paper placed on top of the decoy letter.

Not only had the secret inks and the methods improved, but the Germans had also become ingenious about hiding the fact

AEF, U.S. secret-ink lab

that an agent was carrying secret ink at all. After several German spies were caught red-handed with bottles of secret ink disguised as medicine, German handlers started to impregnate articles of clothing—lingerie, handkerchiefs, socks, scarves, hats, padded shoulders, suspenders, coat collars, ties, and shoelaces—with secret ink.

Now the seesaw battle of wits had tipped back to the Germans. As a result, the British and French desperately started searching for a true universal developer. Although Herbert Yardley spilled a lot of beans in his tell-all book, he did not disclose the composition of the universal developer, saying that it would be unethical to do so. Now, almost a hundred years later, this secret can finally be revealed. Surprisingly, in addition to the secret sources, the method can be found in publications by

French forensic scientists at work during World War I and the following two decades.

Working out of two attic rooms in a gloomy courthouse in Lyon, France, Edmond Locard (1872–1966), a wispy, dark Frenchman with an aquiline nose and a thin black mustache, created the first police forensic science lab in 1910. As a student of the great forensic pioneer Alexandre Lacassagne, he studied medicine and law and became his successor at the University of Lyon. He loved reading Conan Doyle's Sherlock Holmes series when he was a child, and by the time he was an established adult criminalist, he became known as the Sherlock Holmes of France.[40] The author of the seven-volume bible of forensic scientists, the *Traité de criminalistique,* was best known for his principle that "every contact leaves a trace."

Locard's laboratory also contributed to counterespionage activities during World War I, including the detection of invisible ink. One of his staff members used potassium iodide dissolved in hydrochloric acid, combined with iodine, sodium, aluminum chloride, and distilled water (a more dangerous formula added glycerine) as a universal developer.[41]

When this red fluid was brushed over many invisible-ink letters with a cotton swab, it developed the ink. Like the vapor iodine test, it settled on disturbed fibers even if they had been moistened. But it in the end it was only one of the three reagents used to brush over suspicious letters. Even though it was dubbed a universal developer, it was usually used in concert with other reagents or methods.

During World War I another general method emerged that allowed investigators to view invisible-ink letters without disturbing or destroying the letters. Very slowly, spy catchers began to use a quartz lamp—or ultraviolet lamp—to make invisible letters visible. When the letters of a secret-ink message using prisoners' "inks" like urine, saliva, lemon juice, or milk are placed under an ultraviolet lamp they fluoresce, but then turn invisible again without leaving a trace. Investigators could take a photograph of the message before it faded, and it could be used in a court of law. The photographed fluorescent-ink letter could be presented to a jury with dramatic effect.[42]

Stinky Ink

Though British counterintelligence spent a considerable amount of time trying to detect enemy secret ink, MI6, British foreign intelligence, experimented with its own recipes. Some of the new methods were sophisticated, but others used the simple organic methods. Officers were especially interested in experimenting with bodily fluids like blood, saliva, urine, and semen—all readily available substances. An advantage of using bodily fluids is that possession of these substances is not proof of guilt. But the crucial trick was to find a substance that did not react to iodine vapors, the general developer.

Mansfield Cumming, head of MI6, started inquiring about new recipes through colleagues working at London University. In the fall of 1915, Walter Kirke, deputy head of military intelligence at general headquarters, heard some interesting news from Cumming. He told Kirke that he thought "the best invisible ink is semen." Since investigators were anxious to find a naturally occurring substance that was easy to use, "C" was delighted when Deputy Chief Censor F. V. Worthington found that semen did not react to the iodine vapor test (though he did not mention that heat develops it and that it appears under an ultraviolet lamp).

MI6 investigators thought they had solved a great problem, and the men started gleefully experimenting with the new discovery. Obviously, the main way to produce semen at the office was through masturbation. The agent who had discovered the covert use of semen reportedly had to transfer to another department after he was teased so much by other staff members. One officer in Copenhagen took the new discovery so seriously that he "stocked it in a bottle—for his letters stank to high heaven and we had to tell him that a fresh operation was necessary for each letter."[43]

Of course, it wasn't just spies that could use bodily fluids like semen, saliva, blood, and urine. Prisoners were obviously the biggest group of frustrated communicators because they didn't have access to more sophisticated methods. During the world wars prisoners of war were active users of the more primitive methods. In fact, British Censorship intercepted hundreds of letters from POWs. Many of them used milk to write secret messages; other common substances were lemon juice, cobalt chloride, alum, urine, and saliva. The messages described atrocious camp conditions and conditions in their respective countries of captivity, as well as military maneuvers.

Since prisoners were not allowed to write about the conditions in the camps, both sides wrote covert messages about abominable treatment. In 1915, a German wrote home using milk as an invisible ink and said: "Our camp is not of the best. It was formerly a wash house. You wouldn't find such a camp in Germany. The wind whistles through the walls and the rain comes in the roof." They had to sleep on a board with straw on it. There were five hundred prisoners in that camp.[44]

British prisoners also complained about the German conditions. A wounded soldier complained that he and other prisoners had been kicked, whipped, and stoned by sentries and spat on by women and children. They ate bread and water and had to sleep on the floor—and the bread was so hard that a sentry's bayonet broke piercing it.[45]

The United States Enters the
Secret-Ink War

IN THE FALL OF 1916 BRITISH censors opened a letter by a most intriguing spy. His name was George Vaux Bacon and he was an American journalist sent to Britain in September 1916 by German Secret Service officers based in New York City. Bacon was a lanky Minnesota-born young man, with light reddish hair, pale blue eyes, a fair complexion, and a weak receding chin. He wore a pair of wire-rimmed round glasses that made him look intellectual. The journalism occupation was no cover story, though. Bacon really was a writer and journalist who worked for magazines like the film tabloid *Photoplay* and served as the New York City publicity representative of a film production company before the German Secret Service recruited him to become a spy.[1]

British counterintelligence wondered why the Germans hadn't sent any journalists into England yet. Most of the spies they captured had commercial cover. Soon after Bacon arrived in England, censors intercepted his mail—with suspicious underlining—to a contact with a cover address in Holland. Unfortunately for Bacon, the Rotterdam address had recently been placed under censorship surveillance because MI5 had received a tip from MI6 that German intelligence was using it as a cover address. Since Bacon planned to travel to Rotterdam like some of the other

German spies who had been apprehended, counterintelligence decided to shadow him during his stay there. The Bacon investigation in turn led to an "important gang of American spies," as MI5 called them, sent to England from the United States by two German master spies based in New York City—Albert O. Sander and Charles Wunnenberg (Karl Wünnenberg in German). Both of these spy runners had cover jobs in journalism and allegedly worked at the Central Powers Film Company.[2]

Although the United States was still neutral when Wunnenberg and Sander were sent over, it had been supplying the Allies with munitions, war materials, and provisions. The Germans had launched a massive sabotage campaign in the United States focused on the East Coast because they wanted to block delivery of materials to the Allies. They even had chemical specialists based in Hoboken, New Jersey, like Dr. Walter Scheele, who created a cigar bomb. Scheele and his spy ring were implicated in the huge Black Tom explosion in July 1916 that woke up half of New York City and destroyed a major munitions warehouse along the New Jersey shoreline, but Scheele escaped to Cuba.[3]

Soon after the Black Tom incident, Wunnenberg approached Bacon in New York City under the alias "Davis." Wunnenberg asked whether Bacon would be willing to go to England to collect information useful for the German government, including antiaircraft defense, troop movements and morale, and information on new battleships. At first Bacon was concerned about the fate of spies in England, but after the spymasters offered him a liberal expense account and twenty-five pounds a week, Bacon agreed.[4]

Bacon then asked how he would get the messages back to the spymasters in New York City. "Surely the British censor will see them," he said nervously.

"No, no," replied Sander. "I will give you the secret of fooling the censor." That secret turned out to be a new invisible ink

German chemists had developed in order to increase security after the lemon juice spy losses.

After Bacon secured a passport with the pretense that he was going to collect war pictures for the Central Press, he went to see Wunnenberg again, who asked, "Have you got a pair of black woolen socks?"

Bacon looked at Wunnenberg in disbelief and told him he had plenty of socks but nothing in black, so he went down to the store and bought some. When he returned Wunnenberg took out a toothpaste-type tube and said, "Give me your socks." As Bacon watched, Wunnenberg smeared a thick brown paste on the top of the socks.

"There . . . that is a secret ink which the English will never discover," Wunnenberg confidently told Bacon. He then instructed Bacon to soak the top of the socks in water, squeeze them out and use the liquid as secret ink when he wrote letters to Holland. He also told him to use a ballpoint pen—to avoid scratches—and rough paper so that the paper would absorb the ink.[5]

Bacon was now ready and outfitted for his spy mission in England. The day after arriving in London he decided to try out his invisible ink. He "soaked the end of the sock containing the ink in a glass of water, producing a light brown liquid about the color of scotch whiskey." With the secret ink he wrote a long letter about conditions in England to his contact in The Hague. After several weeks he traveled to Holland to meet his spy contacts. He spent the rest of the fall writing several legitimate articles for magazines in New York. During November and December he spent quite a bit of time in Ireland, collecting information on the Irish independence movement for his handlers while eluding the watchers. It was during Bacon's time in Ireland that British counterintelligence made its move. He received a letter asking him to appear at Scotland Yard on a "confidential matter." Originally officials planned only to interview him "in the hope that he could be 'frightened out of the country.'"[6] Sensing that port authorities would have been alerted to prevent any escape attempt, Bacon showed up for the interview on December 9.

Grilled about the people he was writing to in Holland, Bacon claimed not to know that he was dealing with Secret Service personnel. He was, he said, simply trying to sell them some films. The investigators did not believe him and detained him on suspicion of being a German spy.[7]

That suspicion soon turned to confident certainty when they searched his belongings and found incriminating secret-writing materials, including the invisible-ink socks. They found ballpoint pens, a bottle of invisible ink, and rough and unglazed paper. Equally incriminating were the checks and hotel bills from his stay in Holland. They even found the code name "Denis," the Dutch master spy, in one of his notebooks.

The strain began to wear on Bacon during his detention at Brixton Prison, and this produced an unexpected turn of events for the British. In February 1917 Bacon wrote the authorities asking to make a statement. When an officer from Scotland Yard arrived, Bacon told him he would feel better if he were able to make a full confession.

Bacon told Scotland Yard about his recruitment by Sander and Wunnenberg and the method by which he had been instructed to communicate to evade the censors. He insisted, like many spies, that he never intended to pass on any damaging information; he was, he claimed, cooperating with the Germans to get an interesting story on espionage.

There was one thing Bacon did not know: what was needed to develop the secret ink he used. He was simply told how to obtain the ink by wringing out the socks. The British set to work using spectroscopic methods to identify the secret-ink substance. Once they did this, they could try to find an appropriate reagent. Through this analysis, chemists determined that Bacon, as well as other spies, was using Argyrol, the commercial name for a silver salt of a protein mixture sold as a light brown powder soluble in water. Argyrol can also be used as an antibacterial and antiseptic; it was used to combat gonorrhea before sulfa drugs and antibiotics were developed in the 1940s. When investigators confronted Bacon with the bottle of medicine, he claimed he was using it as an antiseptic but did not know that it was impregnated in the socks.

The case of the impregnated socks marked a big change in the Germans' methods. The British and French had found reagents for the simpler secret inks, like revealing iron chloride with potassium ferrocyanide concealed in soap, and lead acetate in perfume that they had come across by searching suspects at border crossings.

But now the Germans had begun to develop methods that did not respond to a single reagent. And the seesaw battle of wits reached a climax. German chemists began to make very dilute solutions, with concentrations ranging from 1:50,000 to 1:500,000. Not only is it hard to detect the secret ink in dilute solutions, but chemical analysis doesn't help because the metal molecule doesn't appear with ordinary developers.

In Britain, the celebrated physicist Thomas Ralph Merton was hired to join the Secret Service after he was rejected for active service because of poor health. Collins provided him with a small amount of liquid, and, using spectroscopy, Merton was able to detect minute amounts of silver from the secret ink impregnated in clothing like Bacon's socks.[8]

Meanwhile, the French also came across cases of clothing impregnated with secret ink—socks and shoelaces—using organic compounds of silver like protargol (similar in composition to the argyrols). It appears as though the French and the British worked together to find a way to develop this extremely dilute secret-ink substance. While Collins headed up the work in the British Postal Censorship chemistry department, the Parisian Department of Judicial Identity, led by Gaston Edmond Bayle (1879–1929), called the "Grand Inquisitor," made major break-throughs in France.[9]

Bayle studied chemistry and worked at the Pasteur Institute in Paris and for the French railway service before joining the French police in January 1915 as a forensic chemist.[10] And just as Collins's investigations sent numerous Germans to the Tower of London to be shot, so too did Bayle and his team have a hand in sending German agents to the dungeons at Vincennes. The Germans blamed treason within their own ranks for these losses, but French Secret Service man Charles Lucieto reprimanded them long after the war for their "clumsiness as chemists." He admonished them for sending their spies to a foreign country equipped with "soap made of potassium ferrocyanide or toilet water that contains lead acetate."[11]

French chemists used electrolysis—a chemical separation method using electricity—to reveal the dilute writing. Apparently, a sheet of paper qualifies as insulating material. If paper with tiny traces of certain metals (the secret-ink substance) is placed in a nascent metal medium—the French used silver nitrate and a reducing agent—the medium is deposited by electrolysis on the trace particles and makes them visible.[12]

Bayle and his team spent months analyzing German secret inks. In addition to working on the dilute silver protein mystery, they spent three months analyzing a dilute secret-ink-impregnated handkerchief officials had seized at the border. They finally discovered the single developer using a catalytic operation (still not declassified) to reveal the secrets.

Bayle was a forensic science hero in France. He was considered brilliant and he solved case after case using science. Since

French intelligence officers were worried about the security of their own secret correspondence, Bayle developed a special secret ink requiring four different reagents used in a set order. He communicated this special secret process to the French War Department in May 1918. No one succeeded in deciphering it, nor has it been made public.[13]

Bayle loved "to unravel a really clever fraud." Even though he was highly regarded internationally because of his skill in unmasking frauds, his victims felt bitter that he found scientific ways to catch them. Some denied any wrongdoing. In 1929, Bayle found that a traveling salesman, a Joseph Philipponet, had produced a fraudulent document to deceive his landlord and obtain money. Seeking revenge, the salesman hung around the police department one morning waiting for Bayle's arrival. As Bayle walked up the steps to his laboratory, Philipponet shot him in the back three times. Bayle swayed, rolled down the stairs, and lay sprawling at the bottom. One last gasp and a mouthful of blood on the floor and he was dead. When the police in the department came to cuff Philipponet, he said: "My document was genuine! What I have done was worth the death of a father of five children!"[14]

Although Bacon failed as a spy, the British thought his secret ink was "first-rate" and the best developed during World War I. Solving the mystery was such a coup that the Postal Censorship department suggested that the king and queen write a secret letter with the substance when they visited the department to thank them for their good work. When the royal couple wrote their names with the celebrated ink, the signatures became brown when developed and chemists fixed the paper with a solution of sodium thiosulfate so the writing would not fade. The framed paper now hangs on a wall in the British counterespionage department.[15]

Collins's testimony about Bacon's secret ink was "directly

responsible" for the death sentence handed down. Bacon was found guilty of charges including collecting information "with the intent to assist the enemy" and "having been in communication with a spy." He was sentenced "to suffer death by hanging," but U.S. officials persuaded the British to commute his sentence to life imprisonment so that he could testify against his New York spymasters. The king approved the commuted sentence.[16]

Meanwhile, in the New York court, Wunnenberg and Sander were each sentenced to two years of penal servitude in March 1917 because of the British tips. Although they initially protested their innocence, they changed their pleas when Assistant District Attorney John C. Knox produced an unexpected piece of evidence. Department of Justice officials had seized all of Sander's papers, including about a dozen blank sheets. Knox began to experiment with the German secret ink in front of the jury. When he dipped the blank sheets of paper in the developing solution, letters began to appear. In a few minutes he had a complete report of Sander communicating with the German master spy in Holland.[17]

Bacon received only a one-year sentence in the United States because the judge "disliked it very much to send such a bright young man to the penitentiary."[18] All three spies served their sentences at the federal penitentiary in Atlanta. Bacon's mug shot as inmate 7097 shows a man who has aged and lost his innocence since his trip to England six months earlier.[19]

America's First Secret-Ink Lab

Because Major Ralph Van Deman, the head of American military intelligence, was worried about German spies using invisible ink in the United States, he asked the Columbia University organic chemist Marston T. Bogert (1868–1954), who was chief

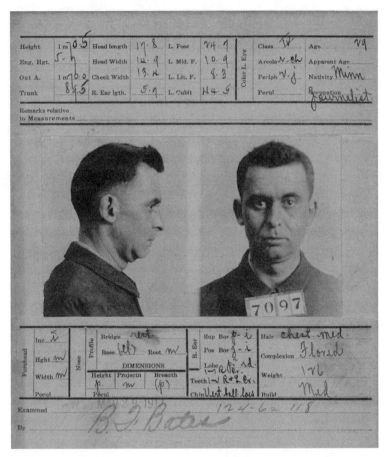

George Vaux Bacon Bertillon card

of the newly founded National Research Council in Washington, D.C., to find an American chemist to help fight the war on secret inks. Bogert contacted the Harvard professor and Nobel Prize–winning chemist Theodore W. Richards to help out. Richards assigned one of his unemployed students, Emmett K. Carver, a twenty-four-year-old Iowan who had just received his Ph.D. under Richards, to do most of the secret-ink work. Carver was sent to New York City to set up a provisional lab in the postal censorship office in November 1917 before the codes and ciphers section of military intelligence started its work in Washington. But before that, they experimented with creating invisible inks as well. Their favorite was blood; when diluted with water, it does not show up on the paper.[20]

The codes and ciphers laboratory was not in full functioning order in Washington until July 1918, but within a few months after Collins's visit and the arrival of detailed plans from the British on how to equip a secret-ink laboratory, Emmett Carver was already feverishly uncovering secret-ink messages by one of the most dangerous spies in American history. Once the lab was officially up and running, Carver—and in his absence Lieutenant A. J. McGrail—possessed a state-of-the-art secret-ink lab. Technicians at the lab examined about two thousand letters a week between its establishment in the summer and February 1, 1919. They detected hidden communication on, and were able to read, fifty letters out of these two thousand. And that's a fairly normal success rate in the field of secret ink![21]

While U.S. military intelligence department 8—the designation MI8 echoed the labels of the British MI system—was struggling to get up to speed on codes, ciphers, shorthand, and secret ink, another U.S. effort had already begun at George Fabyan's famed five hundred–acre private research center, the Riverbank Laboratory in Geneva, Illinois. Fabyan's notoriety rested on his claim that Francis Bacon really wrote William Shakespeare's plays. Researchers were hard at work decoding ciphers in Shakespeare's works to prove this contention. But Fabyan, a successful cotton merchant who also had some inherited wealth, had broad scientific interests, and his private campus also worked on problems in genetics.

Fabyan hired William Friedman, a recent Ph.D. from Cornell, to work in genetics, but soon Friedman became fascinated by ciphers and codes, and he quickly became America's leading cryptologist and cryptanalyst. During the early years of World War I, Riverbank Lab assisted the War Department in deciphering German and Mexican codes and ciphers. In the fall of 1917, they trained more than eighty students in breaking codes and ciphers for the war effort. To teach these students, Friedman started producing a series of technical manuals on cryptography under the rubric of Riverbank Publications.[22]

The Riverbank Laboratory's Department of Ciphers also produced a manual in 1918 on secret ink written by a technician in the laboratory that became very helpful to examiners looking for invisible ink as well as to secret-ink creators. One could imagine that investigators used his manual when searching for secret ink because it is quite thorough. He rightly notes that most secret inks can be revealed when heated or treated with silver nitrate and light. Conversely, he wrote, when creating a secret ink it is of utmost importance to prevent it from being detected or suspected in the first place.[23]

The Blonde Mata Hari?

Bacon was an American sent to Britain by the Germans and caught by British counterintelligence. But the Germans also stationed spies in America (like Wunnenberg and Sander), especially in New York City, as part of their ever-growing secret spy war against America. One of the most dangerous spies in American history was an overweight, handsome blonde woman with a Spanish last name—Madame Maria de Victorica—often dubbed the "beautiful blonde woman from Antwerp" because intelligence officers thought she was "Fräulein Doktor" Schragmüller, head of the Antwerp spy school, and, along with Mata Hari, one of the most famous World War I women spies.

But de Victorica was neither Schragmüller nor the Argentinean that her last name implied. She had, in fact, been born into a Prussian aristocratic family animated by the notion of noblesse oblige. Her education was austere—keeping a stiff upper lip, thinking of money as dirty, and loving one's country were all part of her aristocratic schooling in childhood.[24] Her father, Hans von Kretschmann (1832–99)—a strict, heavyset man with brilliant blue eyes, a fair complexion, golden hair, and a walrus mustache—was a famous general who had fought in the Franco-Prussian War. Although he was not wealthy, when he married Countess Jenny von Gustedt—a granddaughter of Jérôme Bonaparte—he became a Baron.[25]

The couple had two daughters, who were far apart in age and temperament. The older daughter, Lily Braun (1865–1916, née von Kretschmann), became Germany's most famous woman at the turn of the twentieth century. She was an early feminist and socialist and advocated economic freedom for women. As the family rebel, she alienated her father, who disliked her rabble-rousing and public appearances.

Born in 1878 in Posen in eastern Prussia, Maria (also known as Mascha) was thirteen years younger than her storied sister. As a child, she conformed to her parents' values and gave no hint of Lily's unconventional behavior.[26] But she revered and loved her sister, whose philosophy and lifestyle influenced her deeply later in life. She spent a lot of time at her sister's home, where she met leading social reformers and luminaries from the University of Berlin.[27]

While she was at the university, Maria became interested in modern art and met Otto Eckmann, a painter and well-known graphic artist and a leading exponent of the Jugendstil, as Art Nouveau was called in Germany. They married and Maria lived through her husband and catered to his needs. She created a salon of leading artists at her home, just as her sister had done for the reform movement. For their honeymoon they traveled to Buenos Aires and Santiago, Chile, where she met Don Augusto Matte, later Chilean ambassador to Berlin and a banker with interests in a nitrate company. After Eckmann died of tuberculosis four years into the marriage, Maria began to enjoy her independence, despite suffering from a chronic sickness contracted from her consumptive husband. It was during this time that she started using morphine. She traveled the world, learned foreign languages, studied in Germany, obtained several university degrees, and wrote magazine articles and screenplays for silent films.[28]

While studying in Heidelberg she met and married—rather stupidly, she tells us later—her second husband, a Chilean-French medical doctor, Pablo Aliro Montero Siemon. After nine months he had a mental breakdown and claimed to have killed all his patients. Maria left him and traveled to Chile for five years. While she was there she continued to work as a journalist, a career she had started quite successfully in Berlin, writing pro-German and pro-Chilean army news features. She also brokered a deal between her friend Matte's nitrate concern and Germany. It was while she was in Chile that the German Foreign Office offered her the job of "diplomatic agent."[29]

When she returned to Berlin she continued to write for newspapers, becoming the only female newspaper editor in Germany. She also worked for the Bureau of Information and Propaganda, which was soon taken over by the Foreign Office. She even found time to compose propaganda films for the Foreign Office.

In 1913 she agreed to marry José Manuel Victorica, a handsome and mysterious globe-trotting Argentinean gentleman, in Hamburg. They didn't marry out of love. In fact, they hardly knew each other. There is no doubt that they took this step on behalf of the German intelligence office in Hamburg. She now had a legal cover name and was protected because she would be seen as a national of a neutral country.[30]

De Victorica was sent to New York City in January 1917 with

master spy Hermann Wessels, alias Carl Rodiger. She arrived on January 21 at New York harbor on the SS *Bergensfjord* from Sweden, equipped with secret-ink-impregnated scarves and stockings, pens, bottles of ink, and an Argentine passport. Her assignment was to write pro-German articles, to support anti-British Irish nationalists, and to develop a spy ring that would plant bombs on British and Allied ships. She was also tasked with outfitting the spies with the invisible ink she brought with her. After the United States entered the war, the ring planned to sabotage munitions plants.[31]

It wasn't until the fall of 1917 that the British gave Yardley and the censorship department a tip that eventually led to de Victorica's undoing. The information was not about the woman but rather about money that was being sent to a cover address in New York City. When the gumshoes began investigating the addresses, they accidentally discovered chilling letters in secret ink that were then forwarded to Carver's secret-ink lab in the censorship department on Washington Street in Greenwich Village. Since the letters used code names, the mysterious spy could not be identified immediately.

This was the American secret-ink department's first big case. And as they examined, and tried to develop, letter after letter, their eyes widened. Intercepted letters from German couriers, from French and British intelligence, and from the American postmaster were all sent to Carver's desk describing bombing plots. Sometimes he couldn't develop the secret writing because the ink was too old; other times he simply didn't have the right developer. But he hit pay dirt with the so-called Maud letter. The cover letter seemed like an innocuous note describing "Frank's" illness and recovery. In spy letters this often means the spy had been under surveillance but was now safe. This was a tip to examiners that the document might contain a hidden message in secret ink.

Carver's team surrounded him as he tested the letter with reagents that had worked with other German secret-ink letters. They watched anxiously to see the blank paper reveal its secrets as he dipped his brush in the reagent chemical and brushed it across the letter. Nothing appeared. Then he took another brush with another reagent. Carver got excited when faint writing started to appear on the third attempt. "It's written in 'F' ink—here's secret writing," he exclaimed.[32] With the aid of Dr. Juliana Haskell, a German-language and literature instructor at Columbia University on leave for war work at the censorship department, they read the letter. The sender instructed the spy to blow up docks, mercury mines, and factories, and provided six cover addresses in neutral countries.[33]

But the plots did not stop here. As Carver and his staff huddled over letter after letter trying to reveal their secrets, they came across a chilling tale. One letter revealed that a German spy had asked her spy chief to order a plastic altar with holy figures through one of the priests the spy had recruited. The spies planned to hide the new German high explosive tetra in the holy figures of saints. It was unlikely that Censorship would find a priest suspicious for ordering such items.[34]

But the writer of these scary letters was hard to find because she went under a string of aliases, sometimes posing as a French-women named Madame Marie de Vussiere, sometimes operating under her maiden name von Kretschmann or her previous married name. She also kept moving from one fashionable New York City hotel to another. The Madame, who wore mink coats, broad picture hats and gowns, two clusters of diamond rings, and a scarf, eluded them.

Meanwhile the gumshoes—the Bureau of Investigation (forerunner to the Federal Bureau of Investigation)—continued to follow leads on everyone connected with the intercepted letters and cover address. After numerous false leads and missteps, a clue fell into their laps.

On April 3 and 4, 1918, Assistant United States Attorney Benjamin A. Matthews interviewed Wunnenberg and Sander in the Atlanta Federal Penitentiary. During this visit the prisoners confessed further details about the ring in hope of being transferred to a military prison. In addition to exposing Wessels, also known as Roeder, Schroers, or Schrojers, Wunnenberg mentioned that he had met de Victorica briefly at the Knickerbocker Hotel when she arrived in New York City, and that she had been tasked with rousing the Irish against the British. The attorney immediately sent copies of a letter outlining this new information to Attorney General John C. Knox in Washington, the local Bureau of Investigation office in Atlanta, and a special agent from the Department of Justice in Atlanta. The Bureau of Investigation immediately sent a special agent to the Knickerbocker Hotel, but de Victorica had stayed there only briefly after arriving in New York. Investigators went from hotel to hotel in the city, looking for this flamboyant Madame who always had a lot of money and paid her bill up front and in cash.

It was Wunnenberg's confession that led to the arrest of Wessels and de Victorica, not successful investigative work by the Bureau of Investigation as described by Yardley in his 1931 book.[35] Interestingly, de Victorica had warned another agent long before Wunnenberg's prison confessions because she thought "he was a man who, if caught, would tell everything."[36]

According to Yardley, at dusk on the evening of April 18, 1918, the gumshoes followed a sixteen-year-old schoolgirl—it turned out this was de Victorica's personal maid, Miss Margaret Sullivan—carrying a newspaper into the imposing Saint Patrick's Cathedral—the church that takes up the city block on Fifth Avenue between 50th and 51st Streets in New York City. They saw the teenager kneel to pray at pew number 30. When she was finished, she left the newspaper on the seat. The only other worshiper in the cathedral was an elderly gray-haired man who also carried a newspaper. He moved over to the schoolgirl's pew and took her newspaper.

As the man left the cathedral, the secret agents supposedly trailed the newspaper-carrying man in a taxi to Penn Station, where he boarded a train to Long Island and they followed. He allegedly got out at the station for the posh Hotel Nassau on Long Beach, entered the hotel, and sat in the opulent lobby smoking a cigarette for half an hour. When he left, a striking blonde in a black silk dress entered the lobby with a pile of magazines and newspapers. As she nonchalantly flipped through the reading material, she picked up the man's mysterious newspaper and got up to leave. Agents apprehended her and she was arrested on a presidential warrant on April 27, 1918. Her spymaster had slipped twenty thousand dollars of spy money for agents and saboteurs between the pages of the newspaper.[37]

When agent Emma R. Jentzer (the first woman to serve in the Bureau of Investigation; E. R. Jentzer in the reports), together with her husband, Harry J. Jentzer, visited de Victorica in her hotel room, she told them that she had been very sick before coming to America and showed them her scars and sores. Her doctor later reported that she had to have "large quantities of pus" removed from puncture wounds on her body. She had a "morphine hypodermic in her arm" while the Bureau agents were in her room and explained to them that she had to use the morphine all the time or else she wouldn't be able to walk. Emma R. Jentzer noted that she "appeared to be very ill."[38]

When they searched her room, they also found several ballpoint pens (at the time, a rarity outside of spy circles) and two silk scarves or mufflers. When they got to the lab, they tested the scarves and found "F" secret ink in it.[39] Yardley's exposé is part fiction, part fact, highlighting "successes" that never were.

Scientists from Walter Nernst's physical chemistry institute in Berlin had given de Victorica two secret-ink-impregnated silk scarves or mufflers. Chemists there instructed her on how to saturate them in cold water and wring them out. She could use iodine tablets dissolved in vinegar to develop the ink.[40] John Eggert, a noted physical chemist, and Hans Schimank, a physicist and historian of science, worked at the institute during the war and discovered these new secret inks employed by the German navy for several years.[41]

The developer used to reveal de Victorica's invisible ink was no secret to Yardley's ink lab. When they intercepted and read her open letters, she cryptically referred to "kalium iodatum" as a medicine and asked her "friend" whether he liked her "specialty" because it is "far better than his medicine."[42]

Military intelligence considered Madame de Victorica and her ring "the most important" group of spies discovered since America's entry into World War I and the "Maud" letter the most important intelligence they had come across. As a result of the information, they were able to alert the War Industries Board to put a watch on the mercury plants throughout the country.

The information the secret-ink bureau uncovered in the letters also led to other arrests, though few convictions. At first de Victorica provided investigators some real information (such as Wessels's aliases), but she also told a lot of lies. They finally threatened to withhold the morphine she was addicted to if she did not tell the whole story. Her addiction was so severe that she drove to New York City three times a week to pick up supplies. One of her coconspirators, an Irishman named Jeremiah O'Leary, also knew of her addiction, but he thought it worked in a reverse fashion: "The damned Dutch dope, they've got her, and if they don't give her dope, she'll say anything."[43]

De Victorica testified in many trials but was never brought to trial for her own case. When she was thrown into an Ellis Island prison, she mysteriously contracted pneumonia, as did another German spy. Rumors swirled that the German Secret Service had secretly infected her and the other traitor, who had revealed information that sent agents to prison.

Unlike Mata Hari, countless other women spies in France, and the lemon juice spies in England, de Victorica was not shot but merely imprisoned. She died of a second bout of pneumonia on August 12, 1920. She was buried in the Gates of Heaven Cemetery in Kensico, New York, a broken woman. Even though the attorneys John Knox and Ben Matthews "learned very little from her," they attended her funeral service "because of the respect in which" they held her.[44]

A lot of nonsense has been written about Maria von Kretschmann, allegedly the "blonde Mata Hari." The myths began as soon as she was arrested, primarily because she perpetuated them herself. Yardley's entertaining yarn added to the myths. She lied through her teeth and maintained her cover story that she was Argentinean. She even told investigators, who passed this information on to journalists, that she was born there and moved to Germany when she was an adolescent. They swallowed this story hook, line, and sinker. If they had consulted a library and looked up the Franco-Prussian letters her father wrote to her mother that she had edited and published with her sister in 1904, they would have found that she had been born and raised a von Kretschmann in Germany. The myths surrounding von Kretschmann have been perpetuated in books and articles to this very day.[45] The real story of Maria von Kretschmann is even more fascinating than the fiction. As a liberated woman, she accomplished more in literary and political circles than many women have in more supportive environments. Although she may have lived in the shadow of her more famous sister, as the only female editor in German journalism and as a spy who used her wits more than her body, she was a remarkable woman.

Maria von Kretschmann's invisible-ink-impregnated scarf was one piece of evidence of her spy activity, along with the numerous chilling secret-ink letters. But some German spies never got caught. And the concealment methods themselves were often more ingenious than sophisticated chemical formulas.

Not only did the spies impregnate collars, shoelaces, socks, lingerie, and other clothes with secret ink, but some also used the body as a canvas. As a result, innocent travelers were sometimes stripped at a border and sprayed with atomizers to see whether their bodies bore invisible-ink messages. An American consul general and his wife who were searched at the border between Germany and Denmark were indignant when the diplomat was stripped naked. German border guards sprayed his clothes and his body with a chemical solution to search for invisible ink. They found nothing.[46] Spies sometimes even engraved secret messages on toenails. They were revealed with powdered charcoal.[47]

It is probably no exaggeration to say that more happened in the development of secret ink during World War I than during the previous three hundred years. The competition between the Allies and Germany accelerated scientific developments and pushed chemists to develop more ingenious methods for creating, concealing, and revealing invisible ink. It was a perfect storm: great strides had been made in chemistry, and forensic and secret service scientists were on the payrolls of new investigative, intelligence, and counterintelligence agencies. Finally, the war itself, and the battle between the Germans and the Allies, was the final catalyst that accelerated the evolution of secret ink. Secret ink was slowly catching up to its big brother cryptography.

But there is more to the story. Germany had expanded its web of espionage to the eastern and western fronts, to the British Isles, and even to the vast continents of North and South

America. Even though many spies were caught in Britain and America, the agents pushed the diplomatic envelope.[48]

The so-called Zimmermann telegram, in which the German foreign secretary invited Mexico to join the Central Powers should the United States enter the war, has often been credited with precipitating American involvement in World War I. But the secret espionage war taking place behind the scenes in Britain and America had already begun to jostle the Americans from neutrality. When President Woodrow Wilson called for a declaration of war against Germany, he referred to German espionage and sabotage in the United States as well as to the Zimmermann telegram:

> The Prussian autocracy was not and could never be our friend [because] from the very outset of the present war it has filled our unsuspecting communities and even our offices of government with spies and set criminal intrigues everywhere afoot.[49]

Wilson's words came several weeks after the apprehension and arrest of Bacon, Wunnenberg, and Sander. He even pardoned Bacon because he blamed the evil Prussians for ensnaring him in their trap, leading the young man to his transgression. Surely these dramatic events played a role in Wilson's thinking when he declared war on Germany on April 6, 1917.

Visible Nazis

AT ABOUT TEN O'CLOCK on the night of Tuesday, March 18, 1941, two men with foreign accents started arguing about the best way to cross the chaotic streets leading to Times Square in New York. Impatient with his companion, the older man, a tall, swarthy fifty-year-old with black hair and horn-rimmed glasses, darted into the middle of Seventh Avenue against the traffic light. Encountering a heavy current of Times Square traffic, bewildering lights, and honking horns, he turned back when he was halfway across the street. A yellow cab hit him immediately, and he rolled into another passenger car that knocked him unconscious. As crowds swelled around the mangled body, his companion, a short, blond, Nordic-looking man with small close-set eyes and thick glasses, grabbed the victim's brown briefcase. The ruddy-faced blond man's head jerked around to the left and then to the right as he charted a getaway path. As he stepped onto the sidewalk, he hunched his shoulders and fled from the scene of the accident.[1]

Policemen quickly surrounded the injured man and stopped traffic. They lifted him and carried him to a nearby store as they waited for an ambulance. It arrived fifteen minutes later and took him to Saint Vincent's Hospital. The dark-haired man, who was later identified as a Spaniard, Don Julio Lopez Lido, died the next day at 4:30 in the afternoon.[2]

The Times Square accident would have remained a minor incident had the dead man's companion not called the Taft Hotel (now the Michelangelo). A favorite with Nazi spies, the twenty-two-story four-tower high-rise hotel took up a huge corner of Seventh Avenue between 50th and 51st Streets. The hotel's otherwise bland, boxy exterior was brightened by the dazzling marquee of the Roxy Theatre, called the "Cathedral of Motion Pictures." Given its size, the Taft was a good hiding place for Nazi spies. But this would soon change.

The short, blond man who made the phone call was, in fact, Kurt Frederick Ludwig, a Nazi spy. He was worried that the police might find incriminating material about him in Lido's belongings. When the night manager of the Taft asked Ludwig who he was, he panicked and hung up the receiver. The night manager then called the local police to inspect the room for clues about the registered guest, Don Julio Lopez Lido, apparently a Spaniard. After finding clothing without labels and documents written in German rather than Spanish, the police contacted the FBI. The dead man's papers indicated that his real name was Ulrich von der Osten and that his brother lived in Denver. Information from interrogations of von der Osten's brother and help from the British Security Coordination (BSC), located at the Rockefeller Center in New York, revealed that the dead man was in fact a major in German military intelligence and counter-intelligence, the Abwehr.[3]

Von der Osten was a top-level Abwehr case officer who had been based in Spain while handling several important agents in America. He then traveled to Shanghai before arriving in New York City via Hawaii. The FBI succeeded in intercepting a report written by von der Osten (using the alias Konrad) for a Mr. Smith outlining in chilling detail defensive measures around

Pearl Harbor, Hawaii, including maps and photographs. Von der Osten noted, "This will be of interest mostly to our yellow allies."[4]

Meanwhile, British censors stationed in the basement of the palatial pink Princess Hotel in Bermuda continued to intercept more secret-ink messages by someone who signed his name with the mysterious alias Joe K.; the more recent letters were dated March and April, around the time of the accident. Joe K.'s steady stream of letters to cover addresses in neutral Spain and Portugal, as well as in Berlin and Munich, had raised a red flag. Noteworthy was the first cryptic letter intercepted and sent to a Lothar Frederick in Berlin in December 1940, containing sensitive Allied shipping information. The cover address reportedly belonged to Heinrich Himmler, the notorious leader of the SS, Germany's internal security force. The British Bermuda station shared its information with the FBI.[5]

Since the outbreak of World War II, the huge, luxurious Princess Hotel on the Hamilton waterfront had been taken over by the British Imperial Postal and Telegraph Censorship Department. It was dubbed Bletchley-in-the-Tropics after the country house where British code breakers cracked the Nazi Enigma code. The work schedule was grueling and the twelve hundred censors spent most of their waking hours painstakingly analyzing every suspicious letter, telegram, or wire that passed through Bermuda. But for their health and enjoyment the staff also swam, played tennis, or hit a few golf balls. When the numerous women censors, the "censorettes," lined up along the swimming pool in their swimsuits, they looked like a Bermuda resort variation on the Rockettes.[6]

Situated in the midst of the Atlantic Ocean, Bermuda had become a refueling point for Pan American Airlines' Clipper Service to Europe. Called a "flying boat" because it landed in water and resembled a cruise liner, the Boeing 314 had plenty of cargo

space for mail. During the stopover the pilots would take mail to
the Princess Hotel for censorship examination. As a result, Bermuda quickly emerged as Britain's most successful outpost of its
colonial censorship network. During the early years of the war,
scrutiny of the mails was still taboo in America, with its long
hands-off tradition epitomized by the proclamation, "Gentlemen don't read each other's mail." The British, by contrast, had
highly refined censorship and detection techniques developed
during the secret-ink war against the Germans in World War I.

Miss Nadya Gardner, a censorship examiner who became
responsible for the case of Joe K., suspected that a Nazi agent
had written the half-broken English. She came across letters that
often used German cognates with inappropriate meanings in
English—for example cannon (the German *Kanone*) to mean a
gun. Even the clear cover letters were lightly coded with hidden
meanings. The writer, who was posing as a leather businessman,
would often refer to orders that could not be filled because of
financial restrictions, which really meant that he could not fulfill
his case officer's requests.

The young censor with a sixth sense for spotting espionage
activity suspected that the letters also had secret ink embedded
in them. She sent them to Dr. Charles Enrique Dent, head of the
Bermuda Science and Technology department. The test came
back negative. Again she "sent the letters up to the lab, and asked
them to be checked for secret writing." Despite "receiv[ing] perpetually negative reports," she "was convinced they had something on them. . . . I just about drove the lab crazy," she recalled
long after the war. "And they said, 'well, you're not a scientist,
we're the technical people, we've tried every damn thing, there's
nothing on these letters.'"

She pressed the lab to try the iodine vapor test, and they said,
"'well, that's old-fashioned.' I said, 'well, try it.'" When they
finally did use the test, "there it was, secret writing."[7]

The secret-writing substance was later identified as pyramidon (PON), a headache remedy often used by the Abwehr.
This was a peculiar choice since the Germans had used it during
World War I. The British had even found a reagent, iron chloride (as well as the iodine vapor test, though that developed it
only some of the time, and not as well), which developed PON,
but they seem to have forgotten their successes from World War
I. Pyramidon was easily available in German drugstores, and
some American ones, with a prescription.[8]

The mysterious Joe K. wrote hundreds of letters to his Nazi
spy handlers, as well as to his wife in Munich. He usually wrote
the secret writing on the back of a typed letter and often coded
the letter with the acronym PON or left a symbol under the
stamp to indicate that the letter had a secret message in it. Joe
K.'s masters in Spain and Berlin sent him detailed instructions
on using "pon": "All information for us, such as questions about
orders, reports on activities, requests for money or other help,
only in Pon."[9]

Within a four-month period, Joe K. had written two hundred
letters, of which one hundred had been intercepted; in twenty of
the intercepted letters the most important information was embedded in secret writing. Half of the secret-writing letters left
visible clues—either pen scratches or dampening of the paper—
that secret writing had been used.[10] While the Bermuda station
was busy with the minutiae of censorship, the FBI had hit a
dead end in identifying the man who had fled the scene of the
Times Square accident.

An unexpected clue came from the Bermuda censors. They
had intercepted a March 25, 1941, letter from Joe K. describing the Times Square accident to Manuel Alonso (allegedly the
notorious Nazi Heinrich Himmler) in Madrid. The tone of the
letter seemed panicky about the accident, and the letter writer
noted that both drivers were Jews. The basic outline of the inci-

dent was in the undisguised part of the letter. The secret writing disclosed the nationality of the consulate (Spanish) that claimed the dead body, the license plate numbers of the cars, and the name of the hospital where the ambulance had taken von der Osten. Bermuda censors immediately forwarded this information to the FBI, whose investigation immediately took off.[11]

Several clues led the FBI gumshoes to the Nazi spy. Joe K. had often referred to "Uncle Dave and Aunt Loni" in his letters. When he feared that he was under surveillance, he reported to his German masters that Uncle Dave and Aunt Loni had sold the store, meaning he was closing down his spy work in New York City. But it was too late for Joe K. The FBI had found a phone number in von der Osten's papers that belonged to a couple named Dave and Loni Harris. They also found an address belonging to their nephew "Fred Ludwig." Fred's handwriting matched Joe K.'s. They were then able to identify the spy as Kurt Frederick Ludwig.[12]

Instead of arresting Ludwig immediately, the FBI kept him under surveillance for five months to gather more information and evidence about his spy ring and targets. They followed him as he visited docks and harbors in New York City to collect information on shipping and military action. They noted that he drove up and down the East Coast inspecting army installations with his strikingly pretty eighteen-year-old blonde secretary from Maspeth, Queens, Lucy Boehmler. She was the Mata Hari of Ludwig's spy ring. She chatted up army personnel at the bars and sweet-talked her way into secure installations when guards rebuffed Ludwig. In May they drove to Florida to visit naval bases and stopped at factories, army camps, and airfields involved with the manufacture of materials for the war. Boehmler prepared Ludwig's reports and recorded his observations using toothpicks dipped in a solution of pyramidon invisible ink.

In the summer of 1941 another enormous Nazi spy ring was

rolled up and many spies were arrested. Although Ludwig was not apprehended, he drove west fearing that he was under suspicion. The FBI followed him closely on the desolate midwestern roads. He soon noticed a black sedan behind him filled with G-men wearing their classic porkpie hats and smart suits. When both cars stopped at a gas station in Illinois, he walked up to them and said: "I beg your pardon, but why are you following me?" To this remark, an agent identified only as "Bill" in the report replied, "What is the matter with you? Are you crazy?" After that incident the G-men kept their distance. Soon Ludwig ditched the car and got on a bus. The FBI arrested him when he arrived in Seattle on August 23 because they thought he was about to flee the country on a boat for Japan. They found several bottles of pyramidon among his effects. When they asked what it was for, he said that he suffered from chronic headaches. But that explanation was quickly refuted by their discovery of another alleged eye solution used as a developer. They also discovered toothpicks singed brown at the ends where they had been dipped in the headache powder, which had been dissolved in alcohol.[13]

The FBI wanted to know more about the real person who hid behind the agent code name Joe K. They found that he was born in 1903 in Freemont, Ohio, the son of immigrant German parents. He left the United States with his parents at the age of two and returned to their German homeland. He married and had three children. Even though Ludwig made some short visits to the States in the 1920s and 1930s, he did not return for a longer stay until the Abwehr sent him to New York City in March 1940 to form a spy ring. As we shall see shortly, he was not the only German-American recruited by Nazi organizations around the same time.

When Ludwig arrived in New York, he sought out a boardinghouse in the Ridgewood section of Queens, which had become a German colony. Using the cover of a leather salesman, he began to recruit spies from the German-American Bund, an American Nazi organization that tried to promote a positive image of the Nazi government in the United States. One of his eight recruits was Lucy Boehmler, who joined the spy ring for fun and excitement.[14]

Disgruntled that Ludwig never paid her the twenty-five dollars a week he had promised for her services, Boehmler emerged as the star witness at the February 1942 trial in New York. On March 13, all nine spies were found guilty of espionage and conspiracy. Ludwig was initially sent to the Atlanta Federal Penitentiary, along with many other espionage prisoners, to serve a twenty-year sentence, but he was transferred to the high-security Alcatraz Island Prison after a year because he was considered a "troublemaker" and an "anti-American" spy. When Atlanta officials had admitted him to the penitentiary, they found that he needed considerable dental work and new glasses. His spy salary had never been enough to cover his medical expenses, but he was taken care of in prison. When he was finally paroled in 1954, he returned to New York and found work as a guard and attendant at the Morgan Library. With permission of the authorities, he obtained a German passport and boarded the SS *Bremen* on May 11, 1960. He arrived eight days later in Bremerhaven, Germany, never to be heard of again. His conditional release ended in 1961.[15]

Another Nazi spy who wrote copious messages in invisible ink was not so lucky. Based in Havana, Heinz Lüning was tried in a secret Cuban military court, convicted of passing on vital information about Allied shipping, and executed by a firing squad. The beetle-browed, dark, sensuous thirty-two-year-old man was the son of a German father and an Italian mother and could have passed for a Spaniard. Posing as a businessman, Lüning had arrived in Cuba and opened a dress shop a little

Kurt Ludwig's mug shots from Alcatraz

more than a year before he was shot dead. A bungling spy, he had spent only a month in Cuba before the Bermuda station was hot on his trail.[16]

Lüning's life seemed to be a series of fiascos. After his mother died of a serious illness, Lüning failed his classes and dropped out of school. Several years later, when he was eighteen, his father committed suicide and he was left an orphan. He wasn't perceived as very intelligent or knowledgeable about anything, and he spent evenings with friends and "girls and drink." A former friend confided to the FBI that "the only thing he was looking for was girls, girls and more girls, together with drink."[17]

After marrying his stepsister, who came from a prominent Chicago family, Lüning aspired to travel to Honduras with his family, but he couldn't afford the passport fees. When war broke out in Germany, he feared that he would be drafted into Hitler's army and looked for other ways to flee the country. Upon the recommendation of a friend, he joined the Abwehr, attending its Hamburg training school. He wasn't a good student in spy school either, but tried to learn the rudiments of communicating with his handlers using radio, secret ink, and microdots.

Tradecraft officers at the Hamburg Abwehr school trained him to use and disguise three types of secret ink. When Lüning's letters were dated with four digits for the year, like 1942, he used an ink called 1942 that contained one half-ounce of alcohol with ten drops of lemon juice. If the letter was dated with two digits—41—ink 1941 was used, which allegedly contained one half-ounce of urine and ten drops of lemon juice. Finally, the most sophisticated ink used a zinc sulfate, and letters were numbered with a Roman numeral, like VI. The problem with the more sophisticated ink was that it required a lot of time and preparation. Lüning had to soak the paper in zinc sulfate for ten to twelve hours before writing on it. As a result, he used it only three or four times.[18]

Using canaries to protect himself against discovery was probably the most potentially effective ploy Lüning devised during
his short spy career. Even though he often misnumbered his
secret-ink letters or failed to use the right secret ingredient, he
wrote messages on a weekly basis in his tiny room in Havana.
This meant that he had spy paraphernalia lying around that
could compromise him. If someone knocked on the door, he
had the excuse that loose birds were flying around the room and
he needed to secure them before opening the door. He could
then quickly hide his secret-ink materials or radio.[19] But in the
end, it wasn't someone walking into the room that led to his
undoing, but rather the British blanket censorship system stationed in Bermuda. They intercepted one of his first letters in
October 1941.

American censorship claimed Lüning as its first major success,
but in fact, the indefatigable Bermuda station initiated the chase
several weeks before America joined the war. Bermuda sorters
had randomly plucked a letter from bags and bags of mail and
passed it on to the chemists for tests. Lo and behold, like magic,
letters began to appear in a seemingly innocuous business letter
from Havana to neutral Lisbon, where the Abwehr had numerous case officers. The alert Bermuda sorters soon began to find
one or two letters with the same handwriting. The secret-ink
messages chronicled merchant shipping in Cuban waters. By the
time the censors found a letter describing changes at the United
States naval base in Guantánamo, Cuba, the United States had
joined the war, and sorters had intercepted forty-three of forty-
eight secret-ink letters, including one bearing Lüning's address.
On August 31, 1942, Cuban police knocked on his door and
arrested him. Two months later the firing squad shot him in the
head and chest and he dropped dead under the hot Cuban sun.[20]

Ludwig and Lüning were caught because they used invisible
ink extensively. When the Abwehr trained its agents in the art of

using invisible ink, the handlers never imagined that the enemy would discover their supersecret ink formulas; they thought they were invincible. Unlike codes and ciphers that announced their nefarious purpose, agents were led to believe that their secret-ink messages were concealed and invisible. They thought this added a cloak of protection to the dagger of their dangerous deeds. As a result, they wrote as much as they desired in secret ink.

It wasn't just Nadya Gardner's persistence or chemists' skills that allowed Britain to read the copious invisible messages. In the fall of 1940, a German agent code-named Hamlet was sent to Lourenço Marques, the capital of Mozambique, to spy for Nazi Germany. He used pyramidon and two other secret inks code-named Apis (Aspirin) and Beate to send secret messages about enemy shipping back to Berlin.

Hamlet had no problem sending messages using pyramidon, Apis, or Beate. The trouble came when he started receiving instructions in invisible ink. His spymasters had not given him the developer or the formula for the reagent to reveal the secret messages. It happened that the Abwehr had sent another agent to Mozambique who had the developer and formulas. He was instructed to share them with Hamlet, but the man never reached the nation's capital. Desperate to read these messages, Hamlet cabled Berlin, asking for a shipment of developer.

Since it was an emergency, Colonel Hans Piekenbrock, the head of the Abwehr's espionage unit, decided to cable the formulas to Hamlet in his highest-grade cipher through the foreign minister's office. On October 11, 1940, Piekenbrock sent the following urgent message: "Imperative you procure chemicals there since no means to ship them from here." He then attached the developer formula for the pyramidon secret ink: The first part of the compound developer had a solution of 1 percent ferrous chloride in water with 25 percent cooking salt, and the second part contained a solution of 1 percent calcium ferrocyanide in water with 25 percent salt.

Piekenbrock asked the agent at the consulate, "Tell Hamlet to proceed as follows when developing: Equal parts of both solutions (a teaspoon of each, for example) should be mixed shortly before use. The letter should be rubbed over with the solution using a small cotton ball wrapped around a toothpick."[21]

What Piekenbrock didn't know is that British Censorship intercepted this cable. Now the Abwehr's most closely guarded secrets were in British hands. It would take a while before they deciphered the cable and connected the dots.

"Tramp"

Even though the Abwehr was unaware that the British had compromised several of its supersecret invisible-ink formulas, it started to develop a novel and more sophisticated way to conceal a larger amount of material by employing microphotography. Agents felt safer behind the invisibility offered by secret ink than they did with the conspicuousness of codes or ciphers. But invisible ink was tedious for agents when they had to send longer messages. A solution to this problem was to take a picture of top-secret material and reduce the image to the size of a postage stamp or the period at the end of this sentence. The receiver could then read the messages with a compound microscope. And agents could send back more material to headquarters in Germany.

The Abwehr quickly introduced microphotography lessons at its Hamburg spy school. One of the first pupils to learn this novel technique was a German who had emigrated to America in 1922, Wilhelm Georg Debrowski.

Born in 1899 in the grimy industrial town of Mühlheim, near

the Rhine River, Debrowski served in the kaiser's army during World War I when he was just fifteen. Disaffected after the war ended, he began to wander around the world on a ship. After landing in Texas, he adopted the Americanized name of William G. Sebold and joined the torrential stream of Germans who emigrated after World War I. At first he worked for Consolidated Aircraft in California, then he married an American woman from New York City in 1934 and became a naturalized U.S. citizen.[22]

Sebold had suffered from stomach ulcers for many years. He finally had surgery in 1938 and decided to travel back to Mühlheim to convalesce with his German family. He boarded the *Deutschland* ocean liner docked in the New York harbor near Times Squares in February 1939. The journey was uneventful until he was ready to disembark in Hamburg. Two men in dark green suits took his landing card and questioned him about the nature of his job at Consolidated Aircraft; he was then allowed to leave. He would learn only some months later whom these men worked for.

After recovering from his ulcer, Sebold quickly found a job at a local turbine factory and settled into a quiet life. When World War II broke out in September, he decided to stay in Germany even though America was still neutral. Then a mysterious letter arrived from Dr. Otto Gassner, signed "Heil Hitler."

Gassner was a Gestapo official who had read the report on Sebold written by the two men in green suits aboard the *Deutschland*. He invited Sebold to Düsseldorf to discuss an issue of mutual importance for the future of Germany. Feeling secure with his American passport, Sebold ignored the letter.

Then another letter came. This time Gassner threatened Sebold. Sebold wrote back that he was not interested. Gassner didn't give up. The third letter made it clear that the "pressure of the State" would bear upon Sebold if he didn't come to the meeting. Gassner described the clothes Sebold would be wearing at his funeral if he didn't show up.

Then Gassner outmaneuvered Sebold. He looked him up in the local police records and found that he had a criminal record. A few years before he emigrated Debrowski had been involved in smuggling and other felonies. Gassner could now blackmail him into cooperating with Nazi Germany as a spy against the United States.[23]

Like many naturalized American citizens, Sebold felt more loyal to his adopted homeland than some natives, but with the threat of exposure of his criminal past, he agreed to cooperate with the Gestapo, which then handed him over to the Abwehr station so that he could learn how to spy. The Hamburg Abwehr station was responsible for training spies sent to America and was interested in recruiting a German-American spy to set up and run a secret radio station in New York. A Dr. Rankin interviewed Sebold and declared him suitable as a wireless operator.

Sebold desperately thought about how he could get in touch with American authorities. He told Dr. Rankin that he needed to send money to his wife in America while he was being trained for three to four months in Hamburg. Sebold convinced Rankin that he could accomplish this at the U.S. consulate in Cologne. The authorities there reported the incident to the FBI, which would contact him several months later in America.[24]

After Sebold got his affairs in order, the Abwehr sent him to spy school in Hamburg in an ordinary-looking corner building on the Klopstockstrasse, where it trained its spies. The Abwehr also had a pension on the corner where spies could sleep. Sebold learned how to use a Leica camera and how to make microphotographs the size of a postage stamp. While other students were taught sabotage techniques using explosives and poisons, Sebold was trained to make codes, how to build and operate shortwave transmitters, and how to use secret ink.[25]

In turned out that "Dr. Rankin" was, in fact, Hamburg Abwehr chief Dr. Nicolaus Adolf Fritz Ritter. A medium-sized man with slightly blond hair, Ritter was a former career officer and had been a textile executive in the United States before returning to Hamburg to organize air espionage against America and Britain.[26] Ritter was celebrated for his success in setting up the biggest spy ring in the world in America. He gained agents' trust by taking new recruits out for drinks at the Café Alster-Pavillon on the banks of the Binnen-Alster Lake, even putting up his most prized agents at the five-star Hotel Vier Jahreszeiten nearby.

After four months of training, Sebold, now code-named Tramp, was summoned to see the school's director, who interrogated him about secret American installations including the Norden bombsight, a viewing device in an aircraft for aiming bombs.[27]

Ritter planned to have Sebold act as the communications liaison for four Nazi agents based in New York City. In February 1940 Sebold set sail from Genoa, Italy, on the American liner the S.S *Washington,* arriving in New York City eleven days later with a new name and a secret watchcase. The case concealed five lengthy documents that had been microphotographed and reduced to the size of postage stamps. Three of the microphotographs contained detailed instructions for Nazi spies and saboteurs in America and were to be given to Lilly Stein, Everett Roeder, and Fritz Duquesne. The other two were to be kept by Sebold. Sebold's personal microphotographs contained instructions on how to prepare a code to communicate with Hamburg, and the other contained passwords and information on how to contact couriers. According to his new passport, he was now William G. Sawyer.[28]

Nazi comic

As the vast ocean liner entered New York harbor, FBI agents were waiting for him on the docks. After passing through customs, he was then taken to FBI headquarters, and the agents made copies of the microphotographs. The FBI was very interested in tradecraft, especially the way in which the Nazis communicated to their agents. Bureau Director J. Edgar Hoover boasted that the FBI had kept German and Japanese espionage at bay by "constantly uncovering new enemy communication techniques."[29]

Hoover was therefore quite alarmed by Sebold's report about his last day at the Nazi spy school. The principal had told the new recruits in his farewell speech that German agents in North and South America were having trouble communicating back

Nazi comic

home because "the Americans have given us a great deal of trouble," but that the Nazis solved this problem with a solution that began to lace spy thrillers after the war: "Before long we shall be communicating back and forth throughout the world with impunity. I cannot explain the method now but watch out for the dots—*lots and lots of little dots!*"[30]

Sebold had been given microphotographs, but not *"lots and lots of little dots."* Not until the fall of 1941, a year after this first tip, did a British double agent bring more details to the FBI about the microdot, details that could have changed the course of history.

Even though FBI higher-ups were fascinated by the new espionage communication Sebold brought with him, they were wary of him and his story. An agent lived with him, and others watched him during his first two months in New York. Once they trusted him, he settled in the city, and with the Abwehr's money and the FBI's operational instructions, "Sawyer" set up a front office—the "Diesel Research Company"—in the Newsweek Building for meetings with his contacts.[31] The FBI installed two-way mirrors in the office and took over the adjacent room as its hideout to film Sebold's meetings with his Nazi contacts. Sixteen-millimeter motion picture cameras were pointed through the outfitted medicine chest mirrors at a big clock and calendar conveniently placed behind Sebold's desk as a visual time and day stamp. Several large dynamic microphones were hidden around the room to pick up the secret conversations. Hundreds of recordings were made on state-of-the-art Presto Disc Recorders over the next sixteen months.[32]

A transmitter was set up in Centerport, Long Island, in a secluded cabin in order for Sebold to communicate with his spymasters in Hamburg. He then sent the information gathered by Ritter's spy ring in New York City. The messages were in fact sent by a much stronger shortwave radio facility run by the FBI in another secret installation on Long Island.[33]

The most important contact Sawyer made was with the New York spy ring leader himself, Frederick Joubert ("Fritz") Duquesne, a swashbuckling South African who had become disaffected with the British during the Boer War. Duquesne first met Sebold at his favorite beer locale, the Little Casino Bar Restaurant. Jammed among the cafés and retail shops on the sidewalks of East 85th Street in Yorkville, Manhattan, the hardly noticeable Little Casino had become a Nazi spy nest and clearinghouse for passing on information. Duquesne ran thirty-two subagents who collected national defense secrets and passed them on to Germany through shortwave radio, couriers on the

Clipper flying boats, and secret ink. He liked to meet his agents at the Casino.[34]

Outwardly, the Little Casino looked like any of the other immigrant haunts dotting New York City in the 1930s and 1940s: Germans and German-Americans liked to hang out at the bar-restaurant because they could listen to soft German music on the jukebox, drink good German beer on tap, and feast on real German cooking. The easygoing owners, the Eichenlaubs, placidly pumped the beer faucet and dished out Wiener schnitzel for their loyal customers. Until one day in June 1941 they were gone. They had suddenly disappeared, along with other spies sprinkled throughout New York City. They had been targets of a successful FBI swoop that had picked up all of Duquesne's agents, whom contemporaries thought looked "like characters out of a Hitchcock thriller": in addition to the Eichenlaubs, the agents included a draftsman who had inspected the Norden bombsight, an engineer at the Sperry Gyroscope Inc., a steward on the Pan American Clipper, an artist's model, and a host of other Germans or German-Americans with a variety of jobs and professions.[35]

On January 2, 1942, the thirty-three members of the Ritter-Duquesne ring were convicted, leading to a combined sentence of three hundred years in prison. The case was quickly hailed by J. Edgar Hoover as the "greatest spy roundup in American history." And the spy who led federal agents to their quarry was one of their few prized double agents, William G. Sebold, real name Wilhelm Georg Debrowksi, aka William Sawyer and code-named Tramp.

When Piekenbrock showed Ritter the front-page headline in the *New York Times,* the German spy exclaimed, "The pig! The traitor!"

"But Ritter," said Piekenbrock, "by your own principles

TRAMP was not a traitor, not even a spy. He was a man who worked for his new fatherland."[36]

In from the Ocean

The unmasking of the Ludwig and Ritter-Duquesne spy rings in 1941 and an unprecedented number of espionage convictions in the winter of 1942 made it seem as if Nazi spies were invading America.

Another event in the summer of 1942 was soon to send shock waves through the consciousness of the American people. With the entry of the United States into World War II in 1941, Hitler was faced with a formidable economic war machine. To throw a wrench into U.S. war production, the Nazis started plotting and training agents to sabotage factories and railroads in the United States. Although reluctant to act as a taxi for the trained saboteurs, the German navy finally agreed to allow eight Abwehr-trained men to board a submarine on its way to the eastern coast of the United States.

The mastermind behind the sabotage preparations, code-named "Operation Pastorius," was a fleshy, loud, bombastic Nazi named Walter Kappe, who had lived in the United States for twelve years. The operation was named after one of the first German immigrants, Franz Daniel Pastorius, who traveled to Pennsylvania in 1683 and helped develop Germantown, a colony of Mennonite and Quaker families. Kappe sought out Germans who had spent time in America but had a negative experience. He also found Nazis with some American experience so that they could blend into the country. Above all, the Germans had to be handy with explosives and be willing saboteurs.[37]

Not only did Kappe have to recruit the right men for the job, he also had to train them in the skill of effective sabotage and secret communication. An idyllic campus setting at Quenz Lake, on a farm in Brandenburg about forty miles west of Berlin, was the ideal hiding place and training ground for the planned death and maiming spree in the United States.

About a fifteen-minute walk from the Brandenburg train station, the Quenz Lake farm was surrounded by apple orchards, was populated with farm animals, and had a greenhouse for vegetables, a cottage, and a converted two-story barn with stables. But on closer inspection, the farm had been transformed into a working sabotage school. The main farmhouse, a squat, two-story, twelve-room stone building, resembled a country club. The accompanying barn featured a classroom inside with a chemical laboratory above the garage where recruits studied the theory and practice of explosives and learned secret-writing techniques. Recruits learned about laxatives that would turn blue when developed by water strained through cigarette ashes, and their teachers taught them to use Ex-Lax and aspirin as secret-ink materials.[38]

Shortly before their departure on May 20, several saboteurs went to the German High Command for further instructions about how to use invisible ink. They used silk handkerchiefs to write letters to communicate with their masters via the mail drop address in Lisbon; they also wrote to a contact man in New Jersey in invisible writing.[39]

By May 1942 the new recruits were ready for their mission to the United States and boarded the submarine headed toward the East Coast. Shortly after midnight on June 13, 1942, four men emerged from one of the submarine carrying crates with explosives and set them on the beach in Amagansett, Long Island. As a young Coast Guard watchman approached the beach, George Dasch, the ringleader of the group, walked up to meet him with a tall tale that they were lost fishermen. When Dasch realized that the watchman did not believe him, he bribed him with $260. An early recruit, George Dasch had spent twelve years drifting in America as a waiter. Tall and wiry, the thirty-year-old had prematurely gray hair, a gaunt face, and an intelligent loquacity.

Whether because of the beach incident or because he had a change of heart about Nazi Germany, Dasch began to get cold feet about the operation. One of the other saboteurs with whom he shared his change of heart—Ernst Burger—also wanted to withdraw from the operation and take the money and run. Dasch called the FBI's New York bureau the next day with his story but was rebuffed as a "nutcase," even though the FBI had been alerted about the landed saboteurs. A few days later Dasch traveled to Washington, D.C., and called the FBI asking to see J. Edgar Hoover personally. In a matter of hours he provided the FBI with the names of the saboteurs in his group as well as the information that four more saboteurs were to land in Jacksonville, Florida, on June 17.

Dasch also provided Special Agent Duane Traynor with the silk handkerchief containing the names and addresses of the contacts in invisible ink. The only problem was that Dasch did not remember how to develop the secret writing. Traynor sent the handkerchief to the FBI lab, where technicians developed the writing with ammonia fumes (ammonium hydroxide).[40]

At a military commission July 8–31, J. W. Magee, a chemist at the FBI's technical laboratory in Washington, testified about his examination of the handkerchief, the first piece of evidence on exhibit. The attorney general asked the chemist to subject the handkerchief to the ammonia fumes. When he placed the handkerchief over the ammonium hydroxide, like magic, red writing began to appear before the eyes of the jury and the attorney general. The red color and developer meant that the chemical substance was phenolphthalein, an ingredient still used in laxatives.[41]

The eight saboteurs were all sentenced to death, but President

Franklin D. Roosevelt commuted Burger's sentence to life in prison and Dasch's to thirty years in prison because they had provided information that led to the capture of the other six Germans. The remaining saboteurs were electrocuted on August 8, 1942, at the District of Columbia jail in an electric chair dubbed "Sparky."[42]

The execution of the saboteurs marked the end of a highly successful spy-catching offensive by the FBI in America. But this did not stem the flow of Nazi spies to other Allied countries. Unbeknownst to the FBI, German spymasters in Hamburg and Berlin were sending another army of spies to America's sister country, the United Kingdom. And unlike the FBI, British counterintelligence preferred to double such agents, not arrest them.

The Mystery of the Microdot

ON A PLEASANT SUMMER DAY in August 1941, Dusko Popov got out of a taxi at the Park Avenue entrance to the Waldorf-Astoria Hotel. The twenty-nine-year-old Yugoslavian visitor carried a briefcase containing seventy thousand dollars in cash, a vial with white crystals to make invisible ink, and four telegrams concealing eleven microscopic dots that hid information with the potential to change the course of history.[1]

Changing the course of history was the last thing on Popov's mind when he rode the elevator up to his room in the opulent Art Deco hotel. He was ready to see New York City. The British Security Coordination (BSC) minder who had joined him in Bermuda during the refueling stop from Lisbon thought Popov should rest from the twenty-two-hour Pan Am Flying Clipper trip. Instead, he took a refreshing cold shower, ordered a club sandwich from room service, and prepared to stroll down Park Avenue.[2]

When Popov left the hotel, he was not alone. Despite the warm recommendation by MI5, Britain's counterintelligence agency, the FBI, which had been apprised of his arrival several days before, "requested that a very discreet surveillance be maintained concerning [Popov's] activities during his stay" at the Waldorf.[3]

The FBI had good reason to watch Popov, or so they thought. The German intelligence organization, the Abwehr, had sent agent "Ivan" to America to build a new agent network after the loss of Kurt Ludwig and the Ritter-Duquesne ring several months before. The FBI wanted to catch, arrest, and convict any German spies Popov might contact. By contrast, the British, who had turned Popov into a double agent, thought the heart of good counterintelligence was playing the double-cross game discreetly and without publicity. Popov was run by MI5 as part of the scheme to turn Nazi agents to their side. For the FBI he was lowly "Confidential Informant ND-63," a criminal not to be trusted. For MI5 he was star agent Tricyle, who ran two subagents, and for the Abwehr he was prized agent Ivan, who sent valuable information from England to neutral Lisbon, the spy capital of World War II.

Usually code names are meant to conceal an agent's identity. The British, however, gleefully ignored this rule and had fun punning or joking or hinting. Tar Robertson, one of the creators of the double-cross system, initially provided Dusko Popov the code name Skoot as a play on his last name Popov ("pop-off"), because Robertson thought he might leave in a hurry.[4] Some months later Robertson rechristened the prized agent Tricycle because Popov received two subagents and was thus leading the two smaller wheels. But the code name was also a double entendre. By the time the case officer had gotten to know Popov, he realized that he was a playboy who liked to bed two women at a time, a predilection that marked Popov's experience in America as well.

But to court and play with women, Popov needed a fun car. When he strolled down the street on his first day in New York, he came across a car dealership on Broadway and instantly bought a convertible maroon Buick Phaeton with red leather seats and a black sunroof. Within a week, he had moved out of the Waldorf and into 530 Park Avenue, a penthouse apartment suite, and acquired a Chinese manservant.[5]

While the FBI had Popov, the son of a wealthy industrialist from Dubrovnik, under surveillance, American officers from the Office of Naval Intelligence (ONI) and the army's G-2 intelligence division swooped in and interviewed him two days after his arrival. Popov assumed that the men he spoke with came from the FBI, but the G-men did not meet with him until a week later, causing much confusion and many misunderstandings.

After the turf battles were cleared up, the FBI sent the assistant director of its New York office, E. J. Connelley, to meet with Popov on August 18, 1941. During the first meeting Popov said nothing about the contents of the briefcase, nothing about the mysterious microdots, and certainly nothing about a threat to Pearl Harbor, as Popov published in his half-ghostwritten memoirs long after the war in 1974.[6]

According to recently released FBI files that include correspondence and memoranda written at the time, it wasn't until his second meeting the next day with his new case officer Charles Lanman, in a room at the Hotel Lincoln, that Popov provided the FBI with a sample of the secret ink, the code book disguised as the Virginia Woolf novel *Night and Day,* and two files of cablegrams. At this meeting Popov mentioned that four of the cablegrams contained eleven dots. He also provided Lanman with a typewritten copy of the instructions contained in the microdots. In his memorandum on this meeting, Connelley wrote to J. Edgar Hoover that he was "able to examine several of these dots under a microscope and verify Popov's statement that these dots contain instructions in German." A few days later, he showed the intrigued J. Edgar Hoover the microdots.[7]

FBI technicians were quite excited about what they thought was new microdot espionage technology. The microdot was an extremely reduced photographic image of a standard sheet of text and was usually the size of a period at the end of a sentence. Hoover considered it a "masterpiece of espionage." Within a couple of weeks of the FBI meeting with Popov, Hoover sent the White House a letter describing the new uncovered micro-dot communication method: "I thought the President and you [his secretary] might be interested in . . . one of the methods used by the German espionage system in transmitting messages to its agents." Although Hoover attached a copy of some of the contents of the questionnaire hidden in the microdots, the letter emphasized the technical tradecraft aspects of the finding and the fact that the Technical Laboratory had magnified the dots four hundred times in order to read the content. The content ap-pended was part of a questionnaire about American production and delivery of airplanes to Britain and U.S. pilot training. One month later Hoover proudly reported that the FBI had devel-oped an even smaller dot than the Germans. At the time, he dis-played no interest in the actual message hidden in the microdot.[8]

But about one-third of Popov's American questionnaire per-tained to Hawaii and specifically Pearl Harbor. Long after the Japanese attack the following December, this issue became the focus of considerable acrimony and controversy. The Germans had provided Popov with questions about ammunition and mine depots, about anchorages, about a submarine station, and about docks in Hawaii and Pearl Harbor, but Hoover never no-ticed their significance.[9]

According to Popov's version of the story, he quickly became disillusioned with the FBI because it refused to provide him with disinformation to send back to the Abwehr. The Bureau even refused to approve a trip Popov was planning to Hawaii to have a good time with a girlfriend and to collect the informa-tion the Germans wanted concerning Pearl Harbor. As a result, Popov began to live it up. He frequented nightclubs like the celebrity-studded Stork Club (to the dismay of Hoover, who frequented the club himself) and drank and danced late into the night.[10]

Popov took Terry Richardson, a new girlfriend he had met through an Austrian friend, to Florida for a beach trip in his new convertible Buick—with the FBI right behind. The couple took separate rooms because the FBI had threatened to prosecute Popov for breaking the Mann Act, which prohibited interstate travel with a woman for "immoral purposes."[11]

Upon his return to New York City, Popov started to correspond with the Abwehr in secret ink because he still had not been able to set up a radio. In Lisbon his case officer had taught him how to mix a small quantity of the invisible-ink crystals in a wine glass three-quarters full of water. In New York City he mixed pyramidon, the headache remedy, with water to send letters to his German case officer's secretary-girlfriend Elisabeth, using coded language. Apparently, only a few of the letters went through, in part because they were held up at British Censorship in Bermuda.[12]

Fed up with his inability to communicate effectively with his German handlers, Popov flew to Rio de Janeiro in November 1941 to meet with Abwehr officials there. When he arrived no one met him, but he happily checked in to the luxurious white stucco Copacabana Palace Hotel facing the famous Copacabana beach. The first week he lounged on the beach and cruised the bars, but he became impatient with the Abwehr's delay in contacting him and finally walked into the German embassy. Hermann Bohny, the assistant naval attaché who headed German espionage activities, greeted him suspiciously, but after hearing the code word "Elisabeth," he relaxed and took the photos and spy material from Popov and handed him ten thousand dollars.

At the embassy, Popov complained that he was having a hard time communicating. The Abwehr officials agreed that it would be a good idea for him to make his own radio to communicate with Lisbon, but in the meantime they instructed him to send secret-ink letters to a mail drop in Lisbon. When Popov left Rio, the Abwehr spymasters there handed him two microdots pasted on a blank hotel message form with further instructions. He was told that he would receive further messages from Lisbon in microdots pasted under the flaps of envelopes.[13]

Popov returned to New York City aboard the SS *Uruguay*. He was on this voyage on December 7, 1941, when the ship's captain announced to the passengers that the Japanese had attacked Pearl Harbor. When Popov arrived back in New York in mid-December, Charlie Lanman was waiting for him at the dock. Long after the war, Popov wrote that he had been eager to talk about Pearl Harbor but that Lanman wanted to learn more about the wooden trays displaying mounted butterflies that Popov had seen at the German Embassy. But according to the 1941 FBI memorandum, upon arrival of the boat Popov immediately told Lanman about the butterfly trays. There is no mention of Pearl Harbor.[14]

Hoover was convinced the butterfly trays hid secret messages. The FBI had become so obsessed with finding the hidden messages that it stopped all traffic of butterfly trays from Brazil to New York. Hoover scrawled a note in the margin of a report about Popov's Rio trip: "This is vitally important. Get copy of 'dots' so I may see them. See that no more butterfly trays are allowed to enter US at any port." The FBI carted off five hundred butterfly trays they found on Popov's ship to examine them at the laboratory. Lab technicians inspected hundreds of other butterfly trays arriving from Rio de Janeiro over the next year. Despite minute and painstaking analysis, technicians found no hidden messages on the trays and returned them to U.S. Customs. The messages on the microdots, meanwhile, contained more instructions on acquiring information about the composition of a smokeless cartridge powder, about ships, and about gun production and construction.[15]

According to Popov, disillusioned with the FBI in the wake of the attack on Pearl Harbor, he found yet more diversions—all allegedly meant to maintain the playboy image he had cultivated in Europe. During the winter he skied in Sun Valley, Idaho; in the summer he rented a house in Locust Valley, Long Island; in every season he continually juggled his women. After the affair with Terry Richardson ended, he rekindled a passionate affair with the actress Simone Simon, whom he had known in Paris before the war. Of course, Popov's lifestyle cost a lot of money. Soon he exhausted the $70,000 the Abwehr had provided and asked the FBI for loans to pay the bills his playboy lifestyle demanded.

It was clear to everyone involved that Popov's American adventure had been a huge disaster. He had spent $86,000 in nine months and owed the FBI another $17,500. Not only did he leave a trail of debts when he left the United States in October 1942, but the Germans began to think that he was indeed controlled by the enemy.

However, instead of disappearing to England into the safe arms of MI5, Popov flew back to Lisbon and met with his German case officer Ludovico von Karstoff. During grueling interrogations, he convinced the Abwehr that his America trip had been a failure because he hadn't been provided with enough money!

After the war was over, and with 20/20 hindsight, many British counterintelligence officials believed that Popov had brought a critical warning to the United States about the attack on Pearl Harbor in one of those eleven microdots. In 1972 Sir John Masterman, chairman of the double-cross program and an Oxford don who wrote mysteries on the side, published a book in which he declared that the so-called Pearl Harbor questionnaire hidden in the microdots Popov brought with him to America "contained a somber but unregarded warning of the subsequent attack upon Pearl Harbour."[16] Tar Robertson, the hands-on director of the double-cross system, blamed the FBI for this lapse in judgment: "No one ever dreamed Hoover would be such a bloody fool." Yet he conceded that the "mistake we made was not to take the Pearl Harbor information and send it separately to Roosevelt."[17] In fact, there is nothing in the files that would indicate the British even saw the Pearl Harbor questionnaire, one of many, in the world-altering way they described after the war.

At the time no one in British intelligence or counterintelligence called Hoover's, Winston Churchill's, or Franklin D. Roosevelt's attention to any impending attack on Pearl Harbor. In fact, the questionnaire never explicitly refers to such an attack but simply solicits intelligence. It was one of hundreds of questionnaires—many tracking information about harbors, shipping, and docks—that the Abwehr provided to its agents. It was only after the attack on Pearl Harbor that the questionnaire assumed sensational significance. It was also concealed much more cleverly and minutely than the hundreds of other questionnaires British intelligence had already found hidden in film reduced to the size of a postage stamp.

The British and the Microdot

It wasn't just the FBI that was fascinated by the Germans' new communication methods. Both at home and abroad, British Imperial Censorship had searched for, and uncovered, a dizzying array of concealments and secret communication methods. They found microdots hidden on a finger or under a toenail, in tie linings, cuffs, and collars, in jacket linings, on shoulder pads, and in seams, in suitcase locks, clasps, handles, and linings, on frames or lenses of glasses, under stones in jewelry, inside book bindings and gummed flaps of envelopes, in split postcards, in razor blades and wrappers, in fountain pens and penknives, in watches and clocks, and in the "full-stop" and letter "o" of written material.[18]

On August 6, 1941, around the same time as Popov received his dots, Guy Liddell, director of counterintelligence at MI5, wrote in his diary: "RAINBOW has received a letter in Portuguese. One of the full-stops has been found . . . to contain a message of one hundred words. RAINBOW has been warned that he might expect some communication in this form. The process is micro-photography." It turned out that the message referred to formulas for secret ink.[19]

Rainbow (Bernie Kiener), was another double agent like Popov, and he had notified MI5 that his Abwehr case officers alerted him to watch out for lots of dots, just as the Hamburg Abwehr branch school had instructed William Sebold to watch out for "lots and lots of little dots."

A few days later, Liddell lunched with Sir Edwin Herbert, head of U.K. censorship, and brought him back to the office to view the "dot." Herbert was impressed and predicted that it would "revolutionise censorship methods." He was going to make every effort to conceal from the enemy that British counterintelligence had discovered the medium. The information would go to several censorship examiners in Lisbon and Bermuda but to no one else for the time being. British intelligence officers also planned to pressure "the Americans to keep the matter secret."[20]

Another issue troubled Liddell: he suspected that the Americans had kept their knowledge of the microdot from British intelligence. This was a time when U.S. counterintelligence and the fledgling Office of Strategic Services (OSS, the forerunner of the Central Intelligence Agency) still sat at the feet of their more experienced British big brothers. It turns out, though, that the FBI had also viewed the microdots only recently, when Popov brought them to the United States, even though the agent William Sebold had mentioned microphotography earlier.

Rainbow and Tricycle were the first agents to receive communication via, and to notify British counterintelligence of, the new method. But as the British learned more about the method, they noticed that their "most secret source," the Enigma/Ultra intercepts, had referred to "dots" as early as November 1940.

By early September, Liddell was convinced that in the future secret mail communications would be in secret ink and the microdot. The challenge, as always, would be to find the invisible secret communications among the bales and bales of letters and packages passing through mail censorship.

A common misunderstanding at the time, and later, was the blurring of the distinction between microphotography in general and the microdot specifically. Both processes involve reducing a photograph so that it must be viewed through magnification, but the image on a microdot is much smaller than conventional microphotography. Microphotography, a precursor to microfilm, has existed since the nineteenth century. Among the large pool of potential claimants to the invention, John Benjamin Dancer, a British optician and instrument maker, is credited with being the first to develop the process of microphotography, in the 1840s. Soon after photography was invented, he miniaturized an image using a reverse microscope, then he attached a microscopic lens with a focal length of 1.5 inches to the end of a camera to magnify it. Initially, this technique was seen as frivolous and without practical application. Among the texts miniaturized were portraits, monuments, and popular subjects like the Lord's Prayer and novelty items.[21]

During the Franco-Prussian War of 1870–71, microphotographs became a serious form of communication. The Germans had surrounded Paris and cut off all communication channels, whether by road, railroad, or telegraph. The only way to get messages in or out of Paris was via balloons or birds. Since written communication is usually bulky and heavy, a Parisian photographer, René Dagron, came up with the idea of taking micro-photographs of written text and transferring them to collodion (emulsion) positives, which were then stripped from the glass. Carrier pigeons carried microphotograph pellicles measuring four inches square out of besieged Paris; each microphotograph could carry three thousand messages.[22]

But spies didn't seem to use the technique again until around the fall of 1940 when the Abwehr started using the microdot operationally. Although the Abwehr had used photographs of text miniaturized to the size of a postage stamp, not until 1940 did it perfect the technology and reduce the photographs to the size of a dot. It could then masquerade as a period at the end of the sentence or a dot on an i.

But all this leads to another mystery. Who really invented this remarkable technique that had so intrigued and impressed MI5 and the FBI? For many years a Professor Zapp was considered to be the inventor of the microdot because J. Edgar Hoover bestowed this honor upon him in his *Reader's Digest* article on the "Enemy's Masterpiece of Espionage," published in the spring of 1946. In this masterpiece of propaganda, he extolled the Germans' ingenuity, burnished the image of the FBI, and mixed fact with fiction. Hoover even went so far as to claim that the "Balkan playboy [Popov] studied under the famous Professor Zapp, inventor of the microdot process, at the Technical High School in Dresden." In fact, Zapp didn't invent the microdot, and Popov never studied under him, nor was he ever a professor at Dresden Technical College. Hoover's piece is a series of half-truths designed to show off one of the FBI's self-proclaimed success stories. Just as the 1945 film *The House on 92nd Street* fictionalized and aggrandized the Duquesne spy ring, so too did Hoover conflate a variety of true cases and fictionalized others in his "Enemy's Masterpiece of Espionage" to create a crazy-quilt myth; some threads were true, but many patches were false or embellished.

Hoover's mythic story was repeated over and over again until William White, an authority on the history of the microdot, pointed out that in fact, Emanuel Goldberg (1881–1970), a Russian Jewish scientist who emigrated to Germany, invented the microdot. In 1925 Goldberg presented his *mikrat,* as he called the microdot, to the Sixth International Congress of Photography in Paris. The iconic image of his public microdot was a reduced picture of the French photography pioneer Joseph Nicéphore Niépce (1765–1833). The dot sat in the middle of what looks to twenty-first-century eyes like a CD or DVD but was, in fact, a round microscope slide. An arrow points to the dot in the middle of the image, which was only three-hundredths of a millimeter high. The souvenir was packed in a small leather-covered box. Reportedly, Goldberg's microdot could fit "50 complete Bibles in a square inch." But even as early as 1925, it was not spiritual leaders who sought out the microdot. Goldberg was dismayed to hear that his microdot had "found entry in the espionage service of different powers."[23]

Like many people, White thought that Hoover confused the purported inventor with Walter Zapp (1905–2003), the shadowy inventor of the famous Minox subminiature spy camera. Zapp denies working on the microdot process during the war. He claims that he worked on the microscope at AEG in Berlin after he fled his Latvian homeland for Germany in 1940. Zapp mentions in a videotaped monologue that after AEG was bombed in 1943, the company merged with Zeiss Ikon, and thereafter he worked there. He does state that Zeiss Ikon was based in Dresden, home of Hoover's mythical Zapp and the Dresden Institute of Technology. Not much is known about the real Zapp's personal life or what he did during the war. One might speculate that the famous Walter Zapp did contract work for the Sicherheitsdienst, or SD, the Nazi Party's security service, while he worked at Zeiss Ikon, but no wartime documents survive to substantiate the claim.[24]

It turns out that the Zapp story was a distortion based on fact but embellished and twisted. In 1986 the FBI declassified a debriefing of a former German agent named Johannes Rudolf Zuehlsdorff, who said that he had been trained in developing microdots by a microdot camera developer named Kurt Zapp. This Zapp reportedly lived in Leipzig but worked on the microdot apparatus in Dresden. But there does not seem to be a trace of a Kurt Zapp in the remaining Nazi files. It is likely a pseudonym an SD officer used with agents.[25]

The Dresden Technical College worked on military contracts during the war, and a group there was actively involved in extreme reduction microphotography. Kurt Zapp purportedly worked at the Institute for Scientific Photography there and adapted Goldberg's microdot technique for use by secret agents. According to the FBI informant, Kurt Zapp worked on the microdot apparatus, later known as the Zapp Cabinet. Agents like Zuehlsdorff had complained that the apparatus was too large, and Zapp was trying to reduce it to the size of a brick.[26]

Kurt Zapp shuttled between Dresden and Berlin. Zuehlsdorff recalls meeting him at the SD's Berlin office in May 1945 to get a photo I.D. for a secret expedition to Remo, India. In Berlin, Zapp apparently worked at the SD's Foreign Intelligence unit for technical communication support (Group VI F), based in an old gray stone mansion on Delbrückstrasse 6A in the leafy green Berlin suburb Grünewald. The SD had "bought" (without ever paying) the garden-embedded villa from Jewish owners. It had been a training school for the SS before it was taken over by the SD. The technical unit occupied four stories of the building from the basement to the third floor. This unit gained notoriety after the war because it became a factory for producing counterfeit British pounds. In addition to churning out counterfeit money, the group produced fake passports, identity cards, radio transmitters, and, notably, miniature cameras.[27]

Again, Popov never met Zapp, was never trained by him, and most certainly did not study under him in Dresden. He did bring the microdot with him to America in August 1941, as Hoover asserts. According to the FBI version of the story, though, a Bureau technician found the dot by rifling through his belongings, not at the prosaic meeting with his case officer Charlie Lanman. When the laboratory agent held the envelope to light, he saw a "sudden tiny glean." The dot had reflected the light. With telling detail, Hoover describes the "infinite care" the agent took when he placed the tip of a hypodermic needle under the shiny point and lifted it out. When it was magnified two hundred times, its "blood-chilling text" was revealed. The fictional blood-chilling text had nothing to do with the actual text on Pearl Harbor; rather, it described the Nazis' quest for information about atomic energy.[28]

Goldberg, a Jew the Nazis would never acknowledge, invented the first incarnation of a microdot, but German intelligence needed to refine, adapt, and reduce the size of his microdot for espionage purposes. As a result, in 1937, the Heeresewaffenamt, the German Army's ordnance office, sent out a circular in search of bids for work on miniature cameras. One of the firms they contacted was Askania-Werke in Berlin-Friedenau, which specialized in optics and fine mechanics, especially for the armaments industry. Dr. Hans Ammann-Brass was the engineer responsible for the microdot project and demonstrated the system to Dr. Heinrich Beck, a specialist on photographic emulsions and chemistry who headed the microphotography section at the Abwehr's secret communication department, the IG (Geheimsache, literally "secret things"), and who worked with the Heereswaffenamt. The Askania modification followed Goldberg's invention closely but attempted to reduce the size of the dot. Askania used the same glass plate as support for the collodion emulsion. The resulting image was fixed but did not need to be developed. The thin collodion sheet containing the image was then removed from the glass plate, and a modified hypodermic needle was used to separate the microdot from the base.

After Germany invaded Poland in 1939, Ammann-Brass fled Berlin for his native Switzerland and presented a sample of a 450-scale reduction microdot at a 1942 briefing at the Swiss General Command in Bern. After Ammann-Brass left, Zeiss-Ikon, the company Goldberg founded and headed in Dresden, continued further developments on the microdot.[29]

Like many people before them, FBI agents and Hoover were fascinated by this seemingly new technology. It was one of the wonders of technology, a real marvel. Surprisingly, the FBI did not do any research in microphotography journals to verify the information they heard from informants. If they had, they would have found out the true history of the microdot, that a marginalized Jew had invented it, that it had existed since at least 1925, and that Kurt Zapp did not exist.

Despite all these half-truths and misrepresentations, one thing is certain. Communicating is at the heart of successful espionage operations. It *is* important to uncover and neutralize new communication methods. Time and time again throughout history penetrating enemy communication has paralyzed the enemy's espionage system. The achievement would be akin to penetrating a new high-tech digital steganography system that al-Qaeda developed in our own age without the enemy knowing it.

Latin America and the Mexican Microdot Ring

Despite the general failure of the Popov mission, the double agent did bring knowledge of the microdot to America. This lead allowed the FBI to disrupt major German espionage operations in South America, where the technique was widely used. In fact, in June 1940, the FBI had recently set up a new intelligence arm in Latin America, the Special Intelligence Service (SIS), to keep an eye on a large network of German spies in South America, Central America, and parts of the Caribbean. The SIS established its headquarters in a rented space in the RCA Building in Rockefeller Plaza under the cover name of Importers and Exporters Service Company.[30] While the OSS was responsible for wartime intelligence in Europe and Asia, the FBI, though technically limited to counterintelligence and domestic security, stepped in to handle intelligence in Latin America.

The FBI had already set up a liaison relationship with the British Security Coordination, its neighbor in the Rockefeller Plaza building on Fifth Avenue in Manhattan. Despite turf battles and tension with the British, the FBI also worked with the Bermuda Imperial Censorship office to watch out for lots of dots. The FBI installed a point man in Hamilton, Bermuda, who was told when censorship examiners intercepted any dots. Now that the censors knew what to look for, they began to find hundreds of dots. By November 1941, the Bermuda office had been instructed to look and test for microdots in intercepted mail. During the next few months examiners identified "twenty-one letters sprinkled with the deadly dots" in mail passing to and from Mexico. They found one of the most important dots on December 6, 1941. It would help them break up one of Germany's largest and most important spy rings in Mexico.[31]

Hundreds of thousands of expatriate Germans lived in Latin America, and they had become major players in business and economics. They were also well connected socially, especially to high-ranking military and government officials. The Hamburg Abwehrstelle (AST) sent its first resident agent to Latin America in May 1939. At that point the region was still a sleepy outpost. The situation changed dramatically in the spring of 1940, by which time the Wehrmacht had conquered most of western Europe. By the fall Britain proved to be a formidable aerial opponent and the United States began to offer military and intelligence support. The SD, usually responsible for political intelligence, collaborated with the Abwehr in South America. In 1944, the SD assumed the role on its own after Abwehr leader Wilhelm Canaris was executed for his part in a foiled assassination attempt against Hitler (some people believe he was hanged by a piece of piano wire).

When the Abwehr and the SD established operations in South America and Mexico, they needed to communicate with German headquarters. They set up the Orga T, short for technical organization, to handle crucial radio and microphotography communications. Gustav Utzinger, a brilliant young man with a mustache who had just completed his Ph.D. in electronics, was brought in to work for German intelligence after the Telefunken Corporation had hired and then transferred him to Brazil. When he fled Brazil because authorities broke up the spy rings there, Utzinger joined the spy group in Buenos Aires, Argentina, and bought or rented remote farms to hide equipment and send secret messages. The Germans bought 350 chickens, installed an incubator, and planted trees to disguise their activities. They even dug a hole in a chicken coop and buried a transmitter in it.[32]

Orga T also set up a microphotography laboratory in a country house called La Choza (meaning "hut" or "shack") in Bella Vista, near Buenos Aires. A twenty-seven-year-old Austrian offset photolithographer and his wife lived there and churned out large numbers of microdots, often reduced images of American publications. The head of the SD in Argentina, Johannes Sieg-fried Becker, then sent the microdots to Germany through the mail or via a courier.[33]

At first agents relied heavily on clandestine shortwave radio receivers and transmitters called Agentenfunk, or Afu. After the government shut down the German clandestine radio stations operating in Brazil in 1942, agents began to operate a series of secret radio transmitters in Buenos Aires to communicate with German stations in Berlin, Hamburg, and Cologne. This chain of secret transmitters was used to send back information collected from periodicals in the United States and Latin America. Time-sensitive information was condensed and sent in code through the radio, whereas entire articles were microphotographed and sent to Germany through a courier usually traveling to Spain on a ship.[34]

Microphotography allowed German agents to send vast amounts of information to Germany. It was time-consuming and dangerous to send long radio messages to Germany because the messages had to be coded to remain secure. Radio messages were also easy to intercept. Though secret ink disguises itself easily, writing long messages is difficult and time-consuming. The microdot solved problems of information gathering and communication in one stroke. The Weimar dream of condens-ing a library into a suitcase had come true. Now the Nazis con-densed thousands of pages of secret information into dozens of dots.

Condensing information was one attraction of the microdot, but concealing it was another. The miniature photographs could be hidden under stamps, stuck under an envelope flap, jammed into spines of books, and sewn into clothing. Microdots mim-icked hundreds of periods in telegrams, business communica-tions, and love letters. Most important, they disguised chilling messages about blowing up seized ships, provided details of war production, charted ship movements through the Panama Canal, and described the extent of the destruction of U.S. oil stores in the attack on Pearl Harbor in other hidden messages.

Mexico City was the site of one of the largest German espionage rings during the war; it was actually three rings run by the Abwehr stations in Berlin, Hamburg, and Cologne, employing about fifty spies altogether. All three rings ended up primarily communicating through microphotographs. They also used codes, secret inks, and clandestine radios, but they all had to rely more heavily on microphotography after the Mexican government shut down the radio station. German agents primarily received instructions in microdots; they rarely used them to send messages because they were too complicated to produce in the field.

But the Mexico City spy rings were fortunate to have a specially trained agent to produce microdots for them. His name was Joachim Karl Rüge, and he had been a general manager for the Mexico City branch of Körting, a German motor company, since 1933. Reminiscent of the Sebold case, the Abwehr recruited Rüge when he returned to Germany for a business trip in August 1939. As a World War I veteran, he was a patriotic German citizen and agreed to work for the army, but was quickly transferred to the Abwehr. After eight months of technical microdot training he was sent back to Mexico via the Trans-Siberian Railroad, Japan, and Hawaii in July 1940 and was given the job of producing microdots for the Mexico City rings.

When Rüge arrived in Mexico City, he joined the already thriving Abwehr Berlin spy ring headed by Georg Nikolaus, considered one of the most effective leaders in Latin America. In addition to his day job as general manager at the company, Rüge was now Abwehr agent Y2983. Upon his return, Rüge and his wife bought a new house in the fashionable Lomas de Chapultepec district, but the spy salary wasn't enough for their needs and Rüge demanded more. Ordinarily Nikolaus wouldn't have tolerated such presumption, but he needed Rüge, his only microdot man, so he put up with his demands and increased his salary. Rüge then bought a Cadillac sedan. Rüge wasn't the only agent in Mexico to demand luxury. Baron Karl Friedrich Von Schleebrügge, Nikolaus's predecessor, drove a Mercedes-Benz and owned a house and an apartment, all courtesy of the Abwehr. The Abwehr's funding of agents' needs seemed limitless and astronomical. The agents in Mexico didn't need a personal line of credit; everything was on a cash-and-carry basis for them, even though they were already financially stable. It may be that the Abwehr funded them with the SD's counterfeit currency created in Delbrückstrasse 6A and the Sachsenhausen concentration camp.[35]

Georg Nikolaus had a voracious appetite for information. He subscribed to thirty newspapers and technical and economic journals. He was especially interested in U.S. military production. But he didn't have to digest the vast amount of information himself. Several of his agents were fluent in English, and he had them read, translate, and collate relevant information.[36]

In April 1941 the first microdot machine arrived in Mexico. It was included in a film shipment from Agfa, Germany's premier film company, and was sent as cargo via Brazil on the German-controlled Italian airline LATI to an intermediary named Frederick Wilhelm. The microdot machine was listed as an innocuous photo enlarger. As soon as they received the machine from Wilhelm, Rüge and Nikolaus rented an apartment where they could work on making microdots—reducing the information Nikolaus received from his agents. By the end of April, Rüge and Nikolaus sent the microdot messages to mail drops in Brazil, Chile, and Sweden. The letters were either forwarded by mail to the next recipient, usually eastbound to Lisbon or Berlin, or broadcast by radio if the message was short.[37]

April was an important month for Nikolaus's future as a spy for another reason. At the beginning of the month the married Nikolaus (his wife and child were in Germany) dumped his Mexican girlfriend Teresa Quintanilla after an eight-month affair. Even though the blond Nikolaus was not particularly handsome—he was balding and of medium height—Quintanilla was quite taken with him. She thought he was serious about her because he had moved into her apartment. But by the end of the month Nikolaus had found a new German lover. Quintanilla promptly informed Mexican authorities about Nikolaus and identified him as a Nazi agent. She also provided them with the names of some of the ring members and information about the clandestine radio run by a "Don Carlos," along with his phone number. After identifying Don Carlos as Carlos Retelsdorf, they closed down his radio.[38] This made the ring even more dependent on the new microdots. But officials did not go after Nikolaus yet.

By December 6, 1941, the Bermuda censorship office had intercepted some early Eastbound microdot letters postmarked in Mexico. A steady stream of microdots continued to flow from Mexico to Germany until the end of January 1942, a few weeks before Mexican police arrested Nikolaus. Shortly after Nikolaus's arrest, Rüge was detained and sent to a concentration camp in Veracruz. This put a stop to the first series of microdot letters.

Meanwhile, Nikolaus was sent to the United States to be deported to Germany along with 239 other Reichsdeutsche. The Mexicans had agreed to send the Germans to the United States as long as they would be repatriated to Germany via a Swedish exchange ship, the SS *Drottingholm,* in New York and would remain out of FBI jurisdiction. The FBI was eager to detain and interrogate Nikolaus, but it had no proof of espionage. When agents searched him before he boarded the ship, they found six microdots concealed under the tongue of his right shoe. The FBI technical laboratory enlarged the dots and could view the chilling contents: the dots contained copies of blueprints of an escape hatch built into a U.S. submarine. With this evidence the FBI refused to repatriate Nikolaus and held him in a U.S. detention camp for the duration of the war.[39]

Rüge was luckier. He was able to bribe officials in the Mexican Ministry of the Interior and was released from the Veracruz concentration camp. But his problems did not stop with his release. Now the FBI was hot on the trail of a mysterious agent code-named Y2983.

Soon after Nikolaus's detention, thirty more microdots were found in a letter intercepted at the Bermuda station. The letter was addressed to Gusek/Berlin from Y2983 in Mexico City. Nikolaus had made payments to Y2983, but the FBI was not permitted to question Nikolaus because the Department of State kept its word to the Mexicans to detain him simply as an alien. Meanwhile, seven more microdots were intercepted, but the contents were in code. By this time, the United States had entered the war.

William Stephenson, the head of the British Security Coordination (BSC) in New York, was dismayed that the FBI wanted to keep the dots, which would tip off the Germans that their communications had been intercepted. As a result, the FBI released some of the microdots and smudged the rest to make them illegible.

Y2983's anonymity began to unravel when the Bermuda experts intercepted a letter from Europe to a post office box in Mexico City. The BSC worked with the FBI on this case, and with information provided by the Bureau, found out that the post office box belonged to Joachim Rüge. The FBI also found a two-year-old letter in its files from Honolulu addressed to a Clara Rüge on the Schäferstrasse in Berlin-Wanssee—Rüge's mother. Investigators now knew that Y2983 was Joachim Rüge.[40]

The FBI had known about the German spy ring in Mexico since 1941 because the ring's radio operator, Carlos Retelsdorf, had to send his messages via double agent William Sebold's transmitter in Centerport, Long Island, which was operated by the FBI. The Bureau had dubbed the Mexican microscopic dot investigation the Clog Case. With the unmasking of Rüge's identity, the FBI was able to neutralize the Mexican ring.[41]

But Rüge was a cat with nine lives. After the arrest of Nikolaus, Rüge took over his spy job in Mexico City. His ring continued to send microdot messages with the compromised communication system well into 1944. During these two years the SIS used the intercepted letters to determine the identity of the other agents and contacts. Eventually, these leads also uncovered more German agents in Buenos Aires and other South American cities.

After the war was over the United States persuaded Mexican officials to round up and repatriate Germans living in Mexico to their homeland. Twenty-one out of the twenty-four Reichsdeutsche living in Mexico were involved in the Mexican Microdot Case. On August 2, 1946, Mexican police attempted to arrest Rüge, who was now operating a chicken farm in a suburb of Mexico City. When they arrived at his door, he waved a pistol in the air and threatened to kill anyone who tried to arrest him. When a reinforced squad arrived the next day to arrest him, they discovered that he had committed suicide. His note read that he did not want to be shipped back to Germany.[42]

These tales about spies during the war show that Nazis had become highly visible to Anglo-American investigators through their communication methods. Investigators had cracked their codes, intercepted their secret-ink messages, and discovered the microdot. The Nazis thought they had solved the problem of sending a large amount of information secretly when they developed and expanded on Emanuel Goldberg's microdot invention. But as British counterintelligence, the FBI, and South American police penetrated and shut down, or controlled, all-important radio communications, the Nazis became more dependent on the compromised microdot system. In the end, their Latin American operations did not seem to be that important despite the amount of resources they poured into them. Conversely, the FBI created the unusual SIS—really a foreign intelligence operation—in Latin America even though the Bureau was traditionally (and legally) responsible for domestic security and internal counterintelligence. The FBI began to score some successes in breaking up espionage rings as it cooperated more fully with the savvy and experienced British intelligence and counterintelligence operatives. Yet the British and American investigators behind the successes remained invisible to the rest of the world, until now.

Invisible Spy Catchers

ONE OF THOSE INVISIBLE spy catchers worked out of the basement of the Princess resort hotel on the island of Bermuda and scored some of the secret spy war's greatest successes. A reserved man, the slender, bespectacled biochemist shied away from the limelight, especially when it came to talking about his secret war work. Even though he had gone to New York City to provide evidence in the Kurt Ludwig trial, he kept his name out of the American newspapers. He, along with his British Secret Service colleagues, was dismayed about "the unfortunate publicity which these trials were given," which "revealed the methods whereby the F.B.I. and our Censors in Bermuda spotted espionage practices." He was worried that "if the present spy trial in U.S.A. causes the Germans to change their methods then we will have to begin all over again."[1]

This man's name was Charles Enrique Dent (1911–76), and he was the head of British Imperial Censorship's Science and Technology department in Hamilton, Bermuda. He was conceived in Singapore while his British-born father worked there as a government chemist. His Spanish mother returned to her home for his birth, but the family went back to England after a year. His whole career path was a bit unusual, starting with his teenage years, when he dropped out of school to work in a bank.

Charles Enrique Dent

He got back on the education track by working as a laboratory technician and enrolled at Imperial College, London, where he received a B.Sc. in Chemistry and was awarded a Ph.D. in 1934 for his work on copper phthalocyanine (marketed as Monastral Blue). He went on to work for ICI Dyestuffs Group Chemical, where he built on, and patented, his discovery of Monastral Blue, a beautiful bright greenish-blue dye used for inks and printing. A few years later he enrolled as a medical student at University College, London, but the war interrupted his medical studies.

By the mid-1930s Dent had already studied secret writing because he thought it inevitable that England would go to war again, and he could put his chemistry, and his work on dyes, to some use. When he was called up during the war, he initially went to France with the British Expeditionary Force and was in charge of a small mobile laboratory looking for secret writing in army mail. When the Expeditionary Forces retreated, he went back to studying medicine, his new avocation. He was then called up again in late 1940 to head the new laboratory in Bermuda. Dent was the man who orchestrated the numerous secret-ink investigation successes we have already learned about in previous chapters. In the first half of 1942 alone, his group intercepted sixty-three agent communications, leading to twenty-seven convictions.[2]

Stanley Collins, dubbed the "world's greatest secret-ink expert," an accolade he earned because of his work during World War I, remained largely invisible to the public; there is no flesh on top of the secret fame. By contrast, Charles Enrique Dent went on to become a recognized biochemist and medical doctor. After his two-year stint in Bermuda, he returned to England to complete his medical studies. During this time he was sent to the United States again to advise American intelligence on setting up its own secret-ink laboratory. In 1945 he was appointed assistant at University College Hospital Medical School and was sent to the recently liberated concentration camp at Bergen-Belsen to study the use of amino acid mixtures in the treatment of starvation. One year later he was back in New York, studying in Rochester on a Rockefeller Foundation fellowship. During his time in Bermuda, Dent met Margaret Ruth Coad, who also worked in intelligence. He married her in England in 1944, and they had six children.[3]

"The Atlantic Citadel of Censorship"

Dent's wife-to-be was one of more than eight hundred women who worked as censors for British Censorship on Bermuda. But most of them were not as lucky in love as Margaret Coad. The male leaders in British Censorship seemed to consider women the most effective "trappers." As a result, women outnumbered men in Bermuda by a large margin. Aside from the sixth sense these women were supposed to possess, a man wrote in all seriousness that "by some quirk in the law of averages, the girls who shone in this work had well-turned ankles." A British Security Coordination (BSC) memo went on to say: "It was fairly certain that a girl with unshapely legs would make a bad trapper and become a square peg in a round hole. . . . Nobody has discovered what part the leg plays. There is here the basis for some fundamental research."[4] Apparently, no one went on to research male sexism in the British intelligence community!

Bermuda certainly sounded like a plum assignment, with its endless sunshine, palm trees swaying in the wind, elegant tropical gardens, blue skies, and the resortlike atmosphere. It beat dreary London by a long shot. But it turned out to be dull for the "censorettes" because most of the men were married or middle-aged or buried in their work. The "Virgin's Lament" nicely captured women's feelings:

I'm just a girl at MI5
and heading for a virgin's grave
— My legs it was wot got me in—
Still I wait for my bit of sin.[5]

Curiously, *Life* magazine's contemporaneous description of the estimated 800 women in the group (out of 880 people, or so the reporter thought) hardly matched the one by the former intelligence officers: the women described here were apparently elderly, their legs bitten by mosquitoes. To add to the demographic problem, the seven-story palatial pink Princess Hotel had become "a veritable inner fortress of censorship." Countless tourists had stayed in the Princess and enjoyed swimming in the courtyard pool and walking through tropical garden grounds, but now the hotel had become offices for activities largely restricted to censorship.

Most of the censors lived in the Hotel Bermudiana, a big yellow building set on a hill overlooking the Hamilton Harbor. From there they walked to work at the Princess about a quarter of a mile up the road on the shore, a road that was dubbed the "Grumble Road." Despite the griping about the austere conditions, workers organized a choir and a theater company, played table tennis and bridge, and attended dancing classes. One censorette even swam across the bay to work.[6]

Even before Dent, his team, and other officials from British Imperial Censorship had set up shop in sunny Bermuda, British intelligence had started coordinating its efforts from a skyscraper in New York City. When war broke out in Europe in September 1939, England fully mobilized and expanded MI5 and MI6; in 1940 a new liaison office—the innocuous-sounding British Security Coordination, BSC—opened in New York. Thousands of secret agents passed under the Atlas statue on Fifth Avenue before entering the Rockefeller skyscraper to take the elevator to the thirty-fifth- and thirty-sixth-floor offices to meet with their handlers.[7]

William Stephenson, known as the "Quiet Canadian" because he stayed in the background and avoided publicity, headed the BSC and was code-named Intrepid. As an established businessman, Stephenson seemed like an unlikely spymaster, but it was a good cover. Besides, it seemed like many of the new American spymasters were amateurs reeled in from the business or legal professions or from academia. William J. Donovan, a lawyer, headed the American Office of Strategic Service (OSS). More flamboyant, much larger, and more present in public, Donovan was nicknamed "Wild Bill," making Stephenson "Little Bill."

Prime Minister Winston Churchill was fascinated by the cloak-

and-dagger world and authorized Stephenson to "assure aid for Britain, to counter the enemy's subversive plans throughout the Western hemisphere . . . and eventually to bring the United States into the war."[8] The mysterious Canadian businessman Stephenson, whose cover profession was passport control officer, succeeded beyond Churchill's wildest dreams, especially in forging successful cooperative relationships with Donovan and J. Edgar Hoover of the FBI. Moreover, the president of the United States, Franklin D. Roosevelt, helped cement the relationship when he told Stephenson that "there should be the closest possible marriage between the FBI and British Intelligence."[9] And what a marriage it became! With little experience in counterespionage techniques to combat the Germans, the FBI leaned heavily on the British, and this dependency soon led to fruitful cooperation.

Cracking the Kurt Ludwig case in 1941 had been the result of a successful collaboration between the FBI and British Censorship's Bermuda station, brokered by the BSC. When the two agencies put the investigation of the accident in Times Square together with the intercepted Joe K. secret-ink letters, they could identify the spy and his ring. At that point, the United States did not have a censorship organization and relied heavily on British Censorship, especially in Bermuda.

British Imperial Censorship had set up field stations throughout the empire: from the Western Hemisphere to the Middle East and Africa to Australia, British censors were looking for evidence of espionage. Bermuda, Trinidad, and Jamaica were the most prominent Atlantic and Caribbean islands with field stations to intercept postal, telegraph, cable, and radio communications from the Western Hemisphere to war-torn Europe. These paradise islands became "exotic" satellites of the ULTRA project to intercept all German radio communications at Bletchley Park. Even though Bermuda had been a refueling stop for

228 Invisible Spy Catchers

planes and ships since the 1930s, scrutiny of communications hadn't been necessary there until the outbreak of war in Europe. Examining mail and cables on U.S. soil had not been an option before America entered the war because of strict neutrality laws and American sentiments against such an invasion of privacy.[10]

British intelligence was keenly aware that secret communication was Nazi Germany's Achilles' heel. The Nazi war machine relied heavily on communication to keep it oiled and running. While Bletchley Park in England intercepted and decoded wireless messages encoded by Germany's vaunted and seemingly impenetrable Enigma machine, the specialty of the Bermuda "trappers" was intercepting, censoring, and examining mail. Their claim to fame in the world of espionage was finding and reading secret-ink letters and helping to solve the microdot mystery.

Since the British thought the only way to save themselves against the Germans was to decipher all of their communication methods, Stephenson worked hard to enlist secretly the United States' intelligence, counterintelligence, and investigative institutions to aid British interests. Even though Congress supported neutrality and an isolationist stance and the State Department officially banned such cooperation, Franklin D. Roosevelt was supportive of secret liaisons between the BSC and the FBI.

It was no accident that exotic tropical locales like Bermuda, Trinidad, and Jamaica figured prominently in Ian Fleming's James Bond novels. Fleming and other British intelligence officers worked and lived on, or at least passed through, these paradise islands during World War II and conducted some of the most audacious and successful operations from them. During the war the highly romanticized James Bond didn't exist, of course—although Dusko Popov came close and is often considered the model for 007. Fleming took the setting and created an extended black-and-white fantasy, with villains facing off against British heroes, and beautiful women emerging Venus-like from the ocean. Bond sometimes even met the CIA agent Felix Leiter for support. Clearly, these paradise locales are more conducive to holidays than to painstaking work bending over envelopes looking for needles in a haystack, but that is what twelve hundred British censors did during the war.

Like the FBI, Fleming embellished, twisted, transformed, and fictionalized realities, the difference being that he never claimed to be writing anything but fiction. Even though there was no real Commander Bond who took orders from his boss Stephenson in his penthouse suite at the Bermuda Princess Hotel, there was this quiet Canadian British business spymaster Stephenson who often passed through Bermuda (later Fleming and Stephenson both lived in Jamaica). Even though there was no Professor Dent, a villainous scientist who worked with Dr. No against James Bond, there was a Dr. Charles Enrique Dent who led the successful Science and Technology division at the Bermuda Censorship station.

As part of building a good relationship with the FBI, Stephenson arranged special agent training visits to London headquarters for briefings on Nazi espionage methods. He also arranged for a visit to British S.I.S. centers in Latin America to discuss the creation of a similar FBI operation there. Since Stephenson also had close relationships with British Imperial Censorship leaders, it was possible to send an experienced FBI agent to the Bermuda station to learn more about mail opening and closing techniques.

A woman who was a Bermuda expert joined Stephenson's staff in New York City and advised the FBI on how to train and recruit women, since the British thought their high degree of "manual dexterity" and their "well-turned ankles" made them the best candidates. When the FBI assistant director started interviewing women at Washington, D.C., headquarters, some interviewees began to wonder why "an elderly G-man" was inspecting their ankles as part of the preliminary screening. Since they were told the assignment might involve travel to Buenos Aires and Rio de Janeiro, they had to be assured that it wasn't the kind of job they were imagining![11]

Obviously, not all secret-ink investigators had the privilege of traveling to exotic British colonies to carry out their work. While Stanley Collins orchestrated his colonial secret-ink empire from London, MI5 was also working behind the scenes to catch and interrogate German spies by intercepting their communications.

H. V. A. ("Vincent") Briscoe (1888–1961), a professor of inorganic chemistry at Imperial College of Science and Technology, London, was the scientific coordinating officer at MI5 and headed the secret-ink section. Briscoe had already gained experience working with Collins catching German spies during World War I and was brought back during World War II while he continued his duties at Imperial College. He conducted a lot of secret war work in addition to his study of secret ink, including investigations of heavy water and the chemistry of thorium and rare earth metals. He was apparently a "keen motorist" and knew the streets of London better than most cabbies.[12]

Whenever a case officer of a German double agent came across any interesting information about the secret ink the agent received, he passed it on to Briscoe. For example, MI5 finally extracted more information about the secret ink that Nickolay Hansen, a Norwegian coal miner who spied for the Abwehr, hid in his tooth. It turns out Hansen's secret ink was quinine secreted in a small bag in his molar. Hansen's Nazi handlers told him he simply needed to roll the quinine between his thumb and forefinger to create a little invisible-ink pencil. British dentists

Nickolay Hansen secreted secret
ink in his molar. He looks like he
has a toothache.

removed the filling the Nazi doctors had placed over the secret-
ink capsule in order to examine and test it.[13] Like some other
inventive invisible-ink ideas, it doesn't seem very practical, as
the spy has to visit a dentist before using the secret writing.

MI5 passed this unique concealment to Briscoe, who found
it "really quite interesting" and was "avid for any other such
details," for he had just compiled the pamphlet "The Search for
Evidence of Secret Graphic Communication," together with
a colleague. The pamphlet contained general suggestions on
where to look for hiding places, but this was the first time they
actually found a case of a spy hiding secret-ink material in a
tooth.[14]

"Gentlemen Don't Read Each Other's Mail"

At the beginning of World War II, Americans were not ready for the spy catcher's game. The open structure of democratic American society and culture was one barrier. With the ethos that "gentlemen don't read each other's mail," Americans abhorred one of the most effective tools for catching spies—an effective and large organization for intercepting and reading secret communication usually housed in a censorship office, like the one in Britain. The other problem was that no counterespionage structures existed.

As the war in Europe progressed, military circles in the United States saw the need to monitor communications and established a general wartime censorship program in June 1941. The Japanese attack on Pearl Harbor, and the subsequent entry of the United States into World War II, quickly galvanized President Roosevelt into creating an Office for Censorship. Byron Price, a square-faced Hoosier with premature white hair, a soft voice, and piercing blue eyes, became the director of that office in December 1941. As the former executive news editor of Associated Press he understood Americans' abhorrence of censorship and declared that only people who disliked censorship should "ever be permitted to exercise" it.[15]

It is unlikely that all of the fifteen thousand anonymous American censors recruited during World War II disliked the notion of censorship. Whether they approved or not, the "unloved creatures" opened millions of pieces of international mail daily, eavesdropped on countless international phone calls, and "edited" movies, books, radio programs, and films.[16]

Americans did not unanimously embrace censorship, even in time of war. Ernest H. Gruening, governor of the Alaska Territory, started a "teapot tempest" when he appeared before the Senate Judiciary Committee to protest a bill to authorize mail censorship between the United States and its territories. Byron Price apparently soothed the tempest with spy stories about successful interceptions of dangerous messages. Price was fully aware that censorship was only a temporary measure and emphasized this point constantly to his staff. Price emphasized that opening mail was a "criminal offense in peacetime, un-American at any time, a vital necessity in wartime."[17]

Part of Censorship's task was defensive. Thousands of news items seen as damaging to national security were kept out of the press and radio, and potential intelligence items on their way to Berlin or Tokyo—or to a soldier's wife back home—were blackened by the censor's pen. They also left other overt marks with the familiar "opened by censor" sticker and cut out passages from letters. Censorship banned mailings like chess games and crossword puzzles because they could contain hidden messages. This kind of censorship was familiar to most Americans and this is what they were told.[18]

But part of the Censorship office's task also became offensive. And the lesser-known workings of the Censorship department would have outraged Americans if they had become widely known. The United States was slow in developing counterespionage technical capabilities by training chemists and cryptanalysts to uncover secret ink and break codes. Even though the FBI had created a formidable technical laboratory in 1932, its primary crime-fighting responsibilities lay in extortion, kidnapping, and bank robbery cases. The Bureau could offer the skills of handwriting analysts and typewriting experts and slowly expanded its responsibilities to include espionage during the war. In general, the FBI emphasized using a watch list to catch spies, while the new Office of Censorship operated on the premise that it was more effective to test suspicious mail of those who were not being watched.[19]

By July 1942, Censorship had trained a couple of people in

the postal field stations, but not until November 1942 could censors finally work in a functioning laboratory in Washington, D.C., with two chemists and two cryptanalysts.[20]

But the main problem seemed to be that the censorship office did not want its job to be reduced to that of simple sorters. Its officials did not relish the idea of intercepting mail through watch-listing and passing items on to the FBI for technical analysis. This turf battle delayed launching an effective technical counterespionage effort and caused tension between the two agencies.

After gaining a measure of acceptance for censorship in America, the next task for the Office of Censorship became executing it properly. In May 1943, Price determined that the office was not using its counterespionage potential "to its full capability." He established a separate division for the "detection of enemy communications" and for improving liaison efforts with other United States and Allied counterespionage agencies. Price, looking for a good cover name for this new spy-catching unit, toyed with the idea of naming it the Department of Forestry and Grazing. Instead, he settled on the more appropriate but equally unrevealing Technical Operations Division (TOD).[21]

Counterintuitively, the new unit was created in part because of too much secrecy. The intelligence aspect of the mission was not the only element that demanded elaborate secrecy, but since censorship "was a dirty word to most Americans" the office had to hide what it was doing. Leaders thought that the establishment of TOD in August 1943 would resolve some of conflicts in carrying out effective counterespionage, conflicts that had arisen because there was insufficient contact, and insufficient cooperation, between field stations and the central office. As a result, "the secrecy of the laboratory operations was so overemphasized beyond the limits of common sense that misunderstandings arose among station personnel and cooperation was made more difficult." Although Censorship leaders believed in protecting the laboratory, they did not want "an aura of cloistered secrecy . . . to develop." From the time TOD opened until January 1945, a short year and a half, secret-ink laboratory personnel almost tripled, from 190 to 560 people.[22]

The Office of Censorship leadership jealously guarded responsibility for detecting secret communications for espionage purposes. Leaders admired the achievements of British Imperial Censorship in Bermuda and were grateful to them for calling America's attention to Germany's extensive network of espionage agents in the West. Not only was the content of the intercepts significant, but investigative agencies were alarmed by the sophistication of secret-ink and photographic concealment methods. They aspired to duplicate the successful British model. This meant reducing the importance of the watch-list method, which focused on a limited list of suspects, and increasing the sleuthing method—detecting concealed messages.[23]

American investigators had learned their lessons from the cases of caught spies profiled in earlier chapters. While two big cases fell into the FBI's lap—the Sebold/Duquesne spy ring and the saboteurs—all the other cases required sophisticated sleuthing and cooperation with the more seasoned British services and their laboratories in Bermuda.

Colonel Harold R. Shaw, an aggressive U.S. Army reserve officer, was appointed head of TOD to serve the field of counterespionage by combating the use of secret ink and by working to detect codes and ciphers. Shaw had been an irrigation engineer in Honolulu in peacetime and headed the postal censorship station there during the first few months of the war. With his degrees in soil physics and chemistry and hydraulics, he was well qualified to serve as irrigation superintendent of a sugar plantation in Oahu. This also turned out to be a good cover for his new job. Who would suspect an irrigation engineer of

heading up technical censorship! In the fall of 1941 Shaw took
part in an intensive two-month course with other reservists who
became the nucleus of postal censorship. Shaw, who maintained
an interest in military intelligence, was also exposed to technical aspects of censorship through lectures by America's greatest
code breaker, William Friedman.[24]

Dr. Elwood C. Pierce, a University of Maryland prewar
chemistry instructor, headed the technical laboratory, while the
secret-ink unit was run by Dr. Willard Breon, another University of Maryland instructor, and Dr. Jonathan White. They had
joined the Office of Censorship when war broke out and created a manual on the detection of secret inks, trained the personnel, and documented active espionage cases. The two heads of
the secret-ink unit employed four chemists—one single woman
and three married women, one of whom, Ethel T. Pierce, was
Elwood Pierce's wife.[25]

The new TOD was housed within the Office of Censorship
headquarters in the triangle-shaped Federal Trade Commission
(FTC) building on Pennsylvania Avenue near the FBI, the Office of Naval Intelligence, the OSS temporary offices, and the
Connecticut Avenue office of the British Security Coordination.
The two technical branches occupied a windowless guarded
area on the top floor of the FTC. There was so much security
and controlled access to the technical divisions that each branch
had its own door entry.[26]

The same month the TOD was established, U.S. Censorship
called together a counterespionage conference "unparalleled in
history" in Miami, Florida, to learn from their experienced and
successful British friends. About fifty leaders of wartime counterespionage agencies attended this three-day conference in a
heavily guarded office building in downtown Miami. The author of the secret history of the Technical Operations Division
doubted whether a conference of similar scope and significance

(and, one might add, *topic*) had ever taken place in the "history of the world." Striking was the worldwide blanket censorship effort in the Western and Eastern Hemispheres, for British Censorship had indeed become an imperial enterprise. Not only that, but the British saw communication interception as the key method for beating the enemy. Charles de Graz, director of British Censorship, put it this way: "Communications are the life blood of this organization and the first point with which Censorship is brought to face is that Allied communications can as readily be used by the enemy as his own." He thought the Allies could control all communication throughout the world.[27]

The long list of presenters and attendees spanned the globe from South Africa to Canada to secret-ink luminaries like British Imperial Censorship chief technical officer Dr. Stanley Collins and chemist Dr. Charles Dent. American agencies like the FBI and naval intelligence also took part. Collins provided a summary of the history of the use and development of secret ink during World War I. He pointed out that countermeasures developed gradually, and that secret-writing analysis was anticipatory. He thought the most effective weapon for hidden-message detection developed during the early part of the previous war had been the ultraviolet lamp.[28] The British experience in World War I combating German spies and their more recent efforts at the Bermuda station proved to be most useful to the American censors.

At the conference, the British again brought up their preference for using women in sorting and examining mail. The controller in charge of censorship, Charles Watkins-Mence, declared that "nearly every woman is a potential spy catcher. Two thirds of the station in Bermuda is composed of women." The British described a sixth sense that women supposedly possessed. Shapely ankles were not mentioned officially at the conference.[29]

Two months after the Miami conference, in November 1943,

Colonel Shaw and Colonel Norman V. Carlson, chief postal censor, visited British Imperial Censorship's Bermuda station. Carlson's mission was to close down the station. American stations in Miami and Puerto Rico for examination of transatlantic communications between Europe and Latin America were slated to replace the legendary Bermuda station. Shaw was eager to meet the technical staff at the Bermuda laboratory to learn more about the methods and techniques the censors used to catch so many German spies. He was impressed by the Bermuda lab's record of success despite its modest resources and equipment in an improvised lab housed in a carefully guarded wing of the Princess Hotel. Dr. Pierce, chief of the censorship laboratories, also spent several months studying British techniques and the cases developed at the Bermuda laboratory. Initially, a training laboratory was set up at the New York City censorship station to train chemists for the field stations. The lab drew on the U.S. secret-ink experience during World War I and especially on British knowledge before the United States' entry into World War II.[30]

By the time the U.S. Technical Operations Division was founded, the British had intercepted hundreds of letters impregnated with invisible ink and as a result had caught numerous spies. By the end of the war, British Imperial Censorship had intercepted 219 significant secret-writing communications, and U.S. Censorship had intercepted another 120 since its later start, bringing the combined tally to 339 major secret-writing intercepts.[31] These figures show how important secret writing was to German communications and how Anglo-American cooperation short-circuited this essential communication method.

Striking in the short history of the existence of the World War II Technical Operations Division is the flurry of activity during the last year and a half of the war. Secret-ink research was intensive and comprehensive. The Office of Censorship worked together with British and Canadian censor chemists, the FBI, and the Army Signal Corps, but U.S. censors did most of the work themselves or contracted it to industrial laboratories or universities.

The Technical Operations Division sought out interested research partners in industry and universities, concentrating on the East Coast, to get some ideas on how to attack the challenging problem of detecting secret writing, in particular microdots. Bell Laboratories in New York was one of the first industrial labs interested in the work, but the Office of Censorship was more ambitious and searched for further help by approaching the Office of Scientific Research and Development (OSRD), headed by the highly respected Dr. Vannevar Bush. The OSRD superseded the National Defense Research Committee (NDRC), the major World War II research and development organization for defense science outside of the military research laboratories at the Departments of the Navy and Army. Best known for the Manhattan Project, which developed the atomic bomb, the committee also worked on radar, sonar, and the proximity fuse. When the OSRD was created in 1941, the NDRC's alphabetical divisions were turned into nineteen numerical divisions, with such specializations as ballistics research, new missiles, explosives, optics, physics, and war metallurgy. Division 19, headed by H. M. Chadwell, was labeled "miscellaneous weapons."[32]

In January 1944, the director of the Office of Censorship, Byron Price, requested that secret-ink research become a part of "regularly assigned projects through the well coordinated Office of Scientific Research and Development." He was particularly interested in using the laboratories at the OSRD's disposal. Division 19 already developed sabotage weapons for the OSS. Secret-ink research quickly developed under Section 19.1 of Division 19 under Dr. G. A. Richter from the Eastman Kodak Company.[33]

Starting in June 1944, Shaw and his technical aides organized and met with a veritable "brain trust" of chemists and physicists at the boardroom of Harvard University's Malinckrodt Chemical Laboratory at 12 Oxford Street in Cambridge every month. Arthur B. Lamb, a professor of chemistry at Harvard who had assisted Theodore W. Richards in secret-ink research during World War I, presided over the group, which included his Cambridge colleague Robley D. Evans, a physicist who ran the radioactivity laboratory at MIT, and his younger colleague Dr. Sanborn C. Brown, who later became interested in the secret ink used in the American Revolutionary War. Other members of the Secret Ink Defense Committee included Dr. Beverly Clarke from Bell Laboratories, a pioneer in microchemistry, Dr. Ellery Harvey from the General Printing Ink Office, S. E. Eaton from Arthur D. Little in Cambridge, and Dr. W. C. Lothrop, a young chemist who taught at Trinity College in Connecticut and served as an active secretary of the committee.[34]

The California Institute of Technology sent no representative to the East Coast meetings, but it had contracts on this topic with the OSRD, and Linus C. Pauling (1901–94) and his research group played an active role in developing new secret inks as well as defensive measures. Pauling was incredibly prolific in a wide variety of fields, from chemistry to molecular biology. He had the unusual distinction of winning two Nobel Prizes—in chemistry in 1954 for his work on the nature of the chemical bond, and the Peace Prize in 1962 for vigorous antinuclear activism. When he was invited to conduct war work for the OSRD he was a full professor of chemistry, chairman of the Division of Chemistry and Chemical Engineering, and director of the Gates and Crellin Laboratories of Chemistry at Caltech. Although he declined to participate in the atomic bomb project, he worked on dozens of defense contracts for the OSRD. Early in the war he developed innovative and practical devices like an oxygen meter, as well as working on other traditional war work, like research on explosives. He also experimented with, and wrote reports on, secret writing, becoming fluent in the practice of secret-ink creation and detection. To his dying day, he never revealed this secret work. In an interview with his biographer Thomas Hager in 1991, he only hinted at it:

LINUS PAULING: Then I was approached with the problem, how can someone who wants to communicate with the government send a message, let's say in the form of an ordinary letter, a sheet of paper, that no one would be able to decipher.

THOMAS HAGER: An unbreakable code.

LINUS PAULING: Yes. So, I had an idea, and we worked for two or three years on this, and then one of the people with me was disappeared from view. He was taken by the government. So, my secret writing ... and I don't know what happened to that. I still don't say what the method of the secret communication is.[35]

Pauling's most innovative contribution to secret inks drew on his biochemical interests and his work on immunology. Working with three Caltech colleagues, Pauling experimented with a water-soluble polysaccharide gum produced from pneumococcus bacteria, the microbe responsible for pneumonia in humans. When produced as a polysaccharide, the bacteria apparently become nontoxic and chemically inactive but remain active immunologically. The ingenuity of this immunologically based ink lies in its requirement of a specific immunological reagent to develop the invisible writing. The document bearing the secret ink is dipped first into a solution of antibody specific to the polysaccharide, then into a dye solution, in this case a Pontacyl Violet 4 B.[36]

The immunological experiment was part of a broader Caltech research program using living organisms as a secret-ink substance. Pauling developed several other secret-ink combinations, together with some Stanford colleagues, including George Beadle, the Nobel Prize–winning geneticist who, with Edward Tatum, discovered the role of the gene in regulating the chemistry of cells. The Caltech secret-writing contract was one of four—the others were Bell Laboratories, General Printing Ink Company, and Arthur D. Little—that contributed to a still-existing Master Chart of secret inks and developers. This master list is unique in the annals of secret writing because it summarizes most of the secret-writing combinations known during World War II.

Dr. Sanborn C. Brown at MIT was another young physicist who contributed a great deal to the OSRD project. He worked on developing a mass-testing X-ray machine that would detect microdots. The modified X-ray machine was so successful it was installed at the New York Censorship Office in 1944. It could detect microdots at the rate of 4,500–5,000 communications in eight hours. Brown also learned how to make microdots and claimed that his were thinner than those made by the Germans. He went to his grave with this wartime secret.[37]

More chillingly, Professor Robley Evans's radioactivity laboratory at MIT worked on creating and detecting secret ink made with radioactive isotopes. U.S. and British censors speculated that the German enemy was using radioactive isotopes for secret writing, and they wanted to find ways to uncover it. The MIT scientists concluded that a Geiger-Müller counter could detect radioactive secret writing. This discovery was so secret that OSRD leaders did not want the specific detection method to be mentioned in any reports. When Stanley Collins remarked in one of his lectures that the laboratories had devised many new devices for uncovering new secret-writing methods like the use of radioactive isotopes, he simply stated that U.S. Censorship had a "remarkable apparatus" for such a test.[38]

Even though the Geiger counter was invented by Hans Geiger in 1908 and refined by Walther Müller in 1928, apparently it was not widely known as a radioactivity detector until after the war. During the war only a limited number of portable Geiger-Müller counters were available in the United States, apart from those already in use for war work. MIT let the Censorship Office use one of its portable Geiger counters.

MIT scientist Robley Evans believed that in addition to artificial radioactive elements, natural sources like polonium could be used in secret inks. Polonium made headline news in 2006 when ex-KGB spy Alexander Litvinenko died after being poisoned with polonium 210. Unlike most of the well-known artificial radioactive elements like cobalt 57, which are gamma or beta emitters, polonium is an alpha emitter and cannot penetrate objects. In the 1940s polonium was commercially available at "practically any hospital" because it was made from "old radon seeds."[39]

The MIT radiation laboratory also experimented with europium and dysprosium in neutron-activated secret writing.[40] After the war, the British interrogated a former Abwehr secret-writing officer, Hans Otto Haehnle, who told them that he had heard of the use of radioactive isotopes only by the Norwegians during World War I.

Curiously, the intensive research on detecting secret ink and microdots (called "duffs" by the British, "pats" by OSRD scientists) began after the height of the use and detection of secret ink in 1941–42. In fact, microdots started to supplant secret ink for the Germans because dots could hide more information and because the British had uncovered so many of their secret-ink methods.

Much of this work remained research in the lab and was not

developed for application; as a result, the TOD recommended maintaining an active peacetime counterespionage unit. The intensive secret-ink research led to a number of new detection methods, like mass application of liquid reagents, electronic methods, and the mass X-ray method for detecting microdots. Chemists developed several new reagents and improved old ones. They also wrote a comprehensive technical manual with a detailed chart of inks and 190 reagents—the Master Chart.

The Wurlitzer Organ

While scientists in the United States belatedly began work on secret ink, the Allies started to score tactical successes in Europe. British double agents had contributed to fooling the Germans into believing that the Allied invasion in June of 1944 was to occur at Pas de Calais, not Normandy. When German occupying forces fled France after the liberation of Paris in August 1944, their espionage organizations left behind some intriguing instruments. French intelligence stumbled across a device on the first floor of an abandoned building on Boulevard Montparnasse that looked like a piece of furniture. It resembled a souped-up roll-top desk with screens, photographic equipment, ultraviolet lamps, and infrared lights. It was sent to an optical institute for examination, experiments, and an expert report. The French concluded that the basic scientific principle behind the machine was phosphorescence. After a piece of paper was exposed to the ultraviolet lamp, the infrared rays extinguished the ultraviolet-activated phosphorescence. Presumably, it was used in conveyor belt fashion to detect secret ink and contraband.[41]

British intelligence passed this report on to H. V. A. Briscoe for his evaluation. He was totally unimpressed. Although he thought the French had conducted a thorough investigation and come to sound technical conclusions, his verdict was that the "whole thing is completely useless" for the practical needs of counterintelligence detection of secret communication.[42]

The machine the French had found had been covered with dust, disassembled, and clearly not in use. But when Colonel Shaw and his TOD team found similar machines in Munich and Hamburg censorship offices, they were more captivated by the mystery machine than was Briscoe. They quickly dubbed it the "Wurlitzer" because it resembled an organ, and they considered the Hamburg machine a "prize discovery in excellent condition." They were so impressed by the machine that they arranged to have it airlifted to the United States and sent to Cambridge, Massachusetts, for analysis.

The OSRD group at MIT examined the massive machine, took it apart, and put it back together again to determine its function and possible usefulness to the Allies. The mission also found an oxine fuming machine, another standard device at censorship stations. Although Shaw was impressed with the technology, he found German censorship personnel "low caliber": extremely politicized and poorly trained.[43]

The German Foe

Many of the German agents sent to the United States and Britain seemed woefully unprepared for a secret mission. Their secret ink seemed primitive and sloppily used. This was puzzling given the Allies' initial positive assessment of their adversaries' abilities. It turned out that the Abwehr was an inefficient, top-heavy military intelligence organization that employed more than twenty-one thousand staff members in 1941, not including agents and informants.[44]

As the war progressed, the Abwehr and its leader, Wilhelm Canaris, lost credibility in the Nazi hierarchy. Although there were many loyal German citizens in the Abwehr, many of them were opposed to the Nazi regime, or so they later claimed. In April 1944 the suspect Abwehr was folded into the increasingly powerful and much more politicized Reichssicherheitshauptamt (RSHA), headed by Walter Schellenberg. First the Abwehr became a military office within the RSHA, then it was totally absorbed by the security service of the RSHA's foreign intelligence department, the Sicherheitsdients (SD)–Auslands Amt VI. The Abwehr's star fell even further when Canaris was arrested in July 1944 because of his involvement in a plot to assassinate Hitler.[45]

But before the Abwehr's fortunes changed, it had developed a rather large department for secret communications headed by a "small, pudgy chemist," Albert Müller, who had already served as a secret-ink chemist in World War I. Between the two world wars he worked for industry. The Abwehr reactivated him in 1937 with the beginning of rearmament. He was now a fifty-year-old captain.[46] It was in the Müller-led department that new secret-ink methods were developed, microdots manufactured, tiny cameras created, the physics of ultraviolet, infrared, and ultrasound experimented with and applied, and fake passports and money forged. Müller's team also developed mass testing methods for the censorship offices scattered in major German cities.[47]

Most of the Abwehr secret-ink methods and techniques were developed at this special department, called IG—department I, for "foreign intelligence," and G for *Geheim,* "secret." Müller gave assignments to about fifteen staff members.

The Germans found it difficult to develop new secret-ink techniques just as they had during World War I. Many years after World War II, Müller recalled that the subject of secret ink had been virgin territory when World War I broke out, with no existing literature available. His stable of chemists tried hundreds of combinations before they found one that fit all the criteria of a good secret ink: it couldn't be made visible by the common spy-catcher methods like the ultraviolet lamp, iodine fumes, heat, or silver nitrate.[48]

It had been surprising that some of the spies caught in the Americas—like Ludwig in New York City and Lüning in Cuba —had used such low-grade and unsecure inks like pyramidon and an adaptation of lemon juice. It turns out that the Abwehr IG unit had handed out secret ink according to security criteria, depending on how trustworthy an agent was or whether the agent was German. Apparently Ludwig and Lüning did not count as trustworthy agents, or perhaps Ludwig's handlers thought he had been corrupted by his time spent in the United States.

Group I of the security classification of secret ink contained the most secure secret inks of all and was safe against "*all* known censorship reagents." It was given only to Germans. One of the most advanced and intriguing secret-ink methods from this group was code-named Albert and used the so-called carbon paper method. Dr. Josef Mühleisen, a Berlin inventor who developed many of the Abwehr's new secret inks, created it in 1942 after the loss of so many German agents in the United States and Britain. Working at a laboratory set up in Müller's house at Schweinfurthstrasse 57A in Berlin-Dahlem, Mühleisen rubbed a sheet of paper in a circular motion with powdered barbituric acid using a piece of cotton-wool. When Albert began to be used operationally, the Abwehr ordered a paper factory, Fabrik Staffel in Witzenhausen, to impregnate pads of paper with the powder, whose composition was kept secret. An agent could then take a sheet from the prepared pad, place it over the intended cover letter, and write directly on it with pencil not applying much pressure.[49]

Group V ink was not secure against detection. In between those extreme levels were Group IV, which was safe against iodine solution or vapor; Group III, safe against specific reagents; and Group II, safe against all reagents except oxine. This last class was given only to the best agents.

Within each of these groups, the secret-ink methods were given code names. Men's first names were used for inks from an agent to Germany, and women's names identified inks used from Germany to the agents. Kurt Ludwig had used PON, an aqueous pyramidon protected by water after-treatment. This was from Group V, the lowest-security ink. There was also a pyramidon solution dissolved in alcohol called PONAL, belonging to the relatively secure Group IV. This was developed using a ferric chloride potassium ferricyanide solution.

Earlier we came across some toothpick-wielding and matchstick-carrying Nazi spies. German agents never used a stylus or pen, instead using toothpicks or wooden sticks or matchsticks. The matchstick method code-named Josef was classified for use with a Group II ink and was invented by Dr. Josef Mühleisen. The head of the match contained silver nitrate covered by a camouflage coating of real matchstick substance so that it would ignite in the normal way. These matches were manufactured at a factory in the Rhineland. The agent was supposed to scrape away the coating and use the matchstick as a drypoint, but the paper was then dipped and pressed to avoid indentations. This ingenious-sounding method was taken out of circulation because the match did not burn well.[50]

Another matchstick method from Group IV called the Heinrich seemed more successful. The British had already learned about it from their double agent Zigzag (Eddie Chapman) and several other agents during the war. This was also a drypoint method that used a quinine base melted directly on top of the matchstick. It was developed using dilute hydrochloric acid or by transferring it onto photographic paper soaked in dilute sulfuric acid. The letter was then examined under UV light.[51]

It wasn't just the Abwehr that developed new secret inks in

Abwehr IG, Berlin

IG: Head: Office: Officers:	Oberst-Lt. Albert MÜLLER Oberst-Lt. LEONHARDT Rittmeister WUNDERLICH succeeded by Hauptmann KRAUSE Oberregrat. Dr. POTT
IG2:	Dr. Hans Otto HAEHNLE (Chemist) Miss KOPPIN
IG4:	Dr. Heinrich BECK: Microdot Developer Mr. Beck (son)
IG5:	Dr. NOETHLING
IGF	Major Dr. Heinz LUDWIG (Chemist) Dr. Josef MÜHLEISEN (Chemist)
IG2, later IGF	Miss Dr. BRATKE (later SD)(Chemist)
Photographer:	Mr. ROHR

Source: NARA, RG 457, Box 202, VO-63; RG 226. British Interrogation Report of Haehnle, 28.

Nazi Germany. The SD under Dr. Taubeck also developed and used novel secret-writing methods "peculiar to the SD" and unknown to the Abwehr. British and American counterintelligence thought Taubeck was "an able and ingenious enthusiast" even though he was a beginner. They sought to get "every scrap of information or materials which might throw light on the subject." One interesting method was a contact method that anticipated some Cold War methods by many years. It involved using an indicating chemical in a blue-black ink that would transfer invisibly to another sheet of paper through dry transfer. The secret message was then imprinted invisibly on the cover document.[52]

The British also uncovered some fun secret-ink methods in

their Middle East colonies. Apparently a case officer's agents in Spanish Morocco carried their invisible-ink information on their turbans. All they needed to do to read the message was "to rinse the turban with fresh cold water." Other agents in the Middle East concealed their secret ink in chewing gum. The technical experts identified a puttylike substance that could be found in pellets smaller than a quarter of an inch and could be concealed under fingernails and in teeth or hair. One agent hid his supply of ink in his collar tag, in the crotch of his pants, on the seam in the back of his pants, and in the linings of his jacket. The experts determined that the chewing gum ink was made of ammonium vanadate and could be detected by graphite.[53]

Working from the skyscrapers of New York City and the resort hotels of Bermuda, British spy catchers had launched some of the most impressive counterespionage operations in history. They also helped train and co-opt the fledgling U.S. intelligence and counterintelligence organizations and their leaders at a time when the United States was still officially neutral. U.S. cooperation with British intelligence before the United States entered the war violated the Neutrality Act and had to be kept secret from the State Department as well as from the general public.

The FBI scored several very public successes, while the British silently turned Nazi agents to work for their side. Even though some of the Allied successes intercepting secret communication became public during the war, the investigators behind the scenes remained largely invisible. The Allied spy catchers in essence won the battle against their German Abwehr foes who worked from the Hamburg Spy School and the gray stone-faced building along the Tirpitzufer canal at Berlin Central. During the war both sides were kept in the dark about their adversaries' personalities and organizations. Even the spy stories rarely revealed who the people were behind the scenes.

12

Out in the Cold

ON DECEMBER 15, 1981, Wolfgang Reif deposited a letter in a freestanding yellow mailbox in snow-covered Communist East Berlin. The mailbox was quite some distance from his apartment because he did not want anyone to link him with the secret spy letter he just mailed. As he walked away wrapped in a heavy overcoat to protect himself from the subfreezing temperatures, he didn't notice that he was under surveillance by the dreaded secret police, the East German Ministry of State Security (MfS), the Stasi.[1] That was in December; the Stasi's surveillance had begun many months before.

Wolfgang Reif was a diplomat for the Communist German Democratic Republic (GDR) who worked at the country's embassy in Jakarta, Indonesia, in the early 1970s. Soon after his arrival there in 1971, the foreign intelligence arm of the Stasi recruited him to work as an agent, but his handlers found his work "unacceptable," and he tried to break away. While he was in Indonesia he also got involved in a racket with a local, illegally importing automobiles into Jakarta. He abused his position at the embassy by claiming the cars were for official use, then sold them to anyone willing to pay the price, receiving a generous cut from his Indonesian business partner. When Reif began to worry that the CIA might blackmail him if it found out about his

251

lucrative side business, he contacted the resident at the U.S. embassy to preempt a messy situation.[2] That, at least, was his story.

The CIA recruited him at the Hilton Hotel in Jakarta in January 1979; given the code name William, he became the Agency's conduit for confidential information about the East German embassy and the Communist Bloc's foreign policy. His CIA handlers also outfitted him with a tiny microdot reader hidden in a felt-tip pen to read their miniaturized instructions on what Reif later told interrogators was called a "micro-chip," also included in the pen. In addition, he received a writing pad with five pieces of carbon paper so that he could write secret messages with the Cold War's signature "dry transfer method." His CIA handlers told him that he could develop their secret-writing messages by sprinkling the paper with dark powder, like cinnamon, blue chalk, or ashes from wood.[3]

Meanwhile, Reif's personal problems snowballed. His marriage was on the rocks, he drank to excess, and his professional life suffered. He thought of defecting to West Germany, and he let his problems be known (under the influence of alcohol, he stated later) to West German colleagues, who informed the Federal Intelligence Service of the Federal Republic of Germany. West German intelligence took the bait and recruited him in Jakarta in April 1980 under the code name Bernhard. In between, though, his marriage improved, and he decided to return to East Germany. He dropped the CIA and began to work only for the West Germans, passing on information in exchange for money. He returned to East Germany in the spring of 1981. And if working for this bewildering number of intelligence agencies wasn't enough, he continued to do some translating for East German foreign intelligence, thus making him a triple agent.

In May 1981, the Stasi's mail searchers intercepted a letter in which the handwriting of the sender did not match that of the letter writer. With their new forensic machine to detect invisible impressions, the Hungarian-named Nyom (Clue), similar to the West's Electrostatic Detection Apparatus, ESDA, they found a number of other suspicious characteristics, such as inconsistent writing pressure, that led them to believe the letter involved intelligence activity. The mail interceptors passed it on to the technical services division's secret-writing detection department.

The technical service chemists had recently developed another new magical technique for uncovering secret writing, using specially prepared paper that developed the secret writing when the paper came into contact with the spy's secret-writing letter. The ingenious chemists soon uncovered the secret message and found out who the spy was through handwriting analysis and by comparing everyone's handwriting in the building to the writing on the letter. The Stasi then placed Reif under surveillance and observed him deposit the letter into the yellow mailbox on that fateful cold and snowy December 1981 day. It was addressed to a well-known West German intelligence cover address and was one of those prewritten letters that the Stasi knew as the hallmark of the West Germans' secret-writing tradecraft.

When Stasi agents searched Reif's apartment they found all his CIA paraphernalia, along with ten prewritten letters, three pads of paper with 195 carbon sheets, and instructions on how to write invisible-ink letters. They even found many letters written in invisible ink. This was all evidence of spy activity.[4]

The Stasi questioned Reif fiercely about the CIA's communication methods. Investigators were particularly interested in the miniature microdot reader and the "micro-chip" (microdot) in the felt-tip pen. He described the concealment of the reader and pen in great detail, noting the length and width. Markus Wolf, head of East German foreign intelligence, made a documentary film of the trial proceedings. The film featured these communication methods and focused on enemy secret service attacks against foreign embassies and East German citizens living in capitalist countries. Reif was forced to talk in great detail about his recruitment.[5]

Reif was one of several dozen spies who were caught because of East Germany's tightly woven net of spy catchers. The mail interception department took a leading role in capturing him and worked closely with the spy catchers' department. His case is just one of many that illustrate the amazing successes that mail interceptors scored in the Communist Eastern Bloc that led to uncovering secret writing. Other Eastern Bloc countries and the KGB itself—the Soviet Union's state security committee and main platform for espionage—were also quite successful in catching spies because of their suffocating surveillance and mail censorship.

Unlike the Stasi and the KGB, the Western democracies were rarely successful in catching spies through mail interception and blanket surveillance. Though secret communication methods were similar in totalitarian and democratic regimes, by their very nature totalitarian regimes had an advantage when it came to intercepting secret messages because of more extensive nets of mail censorship. While many agents who worked for the Western Bloc were caught through interception of their secret communication, when agents who worked for the East were caught, it was usually because of tips from a defector, not because their mail was intercepted and the secret communication deciphered and read by Western agents. As a result, many Western spies working against the Eastern Bloc were left out in the cold after communicating.

In one of his more lucid moments, the bombastic Erich Mielke, head of the East German Ministry of State Security, insightfully pointed out to his staff members that communication is the most vulnerable and most important part of espionage operations, "the nerve center and simultaneously the most sensitive area of attack of operational work."[6]

Imagine for a minute not being able to communicate. In our own age of Internet volubility it would be frustrating not being able to connect, but for spies it is absolutely essential: no communication, no intelligence transmission or rendezvous arrangement. Of course, the problem for most intelligence agencies is providing safe, secure, and secret communication for its agents. While codes and ciphers offer secure communication by scrambling messages, hidden writing offers secret communication by making it invisible and therefore not as easily detectable.

Though some people associate invisible ink with child's play, during the Cold War it continued to be serious business as it had been in the two world wars. But rather than call the technique the childish-sounding "invisible ink," secret services began to use the term "secret writing," or SW, a characterization that includes most steganographic techniques, like microphotography as well as secret ink.

Thousands of years ago, the use of covered or secret writing had changed the course of history. Unlike its big brother, cryptography, which visibly scored several "changing the course of history" successes in the twentieth century, with cases like the Zimmermann telegram and the breaking of the Enigma code, steganography's breakthroughs were seemingly more modest. Secret ink and the microdot did indeed provide messages that helped save D-Day, but codes and ciphers played a more visible role in the everyday life of embassies and military units throughout the twentieth century.

Aside from helping to convict enemy spies and decipher their secret messages, or successfully hide friendly agents' secret messages, secret-writing creation and detection reached a new phase in its scientific evolution during the Cold War. The seesaw secret-ink struggle of World War II became a veritable arms race between the Eastern and Western Blocs that changed the course of the science of communication secrecy.

Innovations in SW were made possible by new developments in chemistry, and the Cold War context accelerated the scientific developments out of necessity. In a sense, it was World War I redux, but with more sophisticated science and a longer period to develop it.

During the early 1950s both sides seemed to have had fairly primitive secret-writing techniques. In the Eastern Bloc and the Soviet Union dead drops in cemeteries were standard means of communicating from agent to handler. Operatives developed mundanely camouflaged hiding places in the cemeteries, like a hollowed-out tree trunk or bolts in the ground. Messages were then left in the dead drop. The CIA used dead freeze-dried rats as concealment devices. When cats started eating their dead rats, the CIA added Tabasco sauce to deter them.[7] In the early years, Eastern Bloc services still used the post office to send messages to cover or accommodation addresses.

Early on, both sides primarily used the so-called wet systems —invisible writing ink from a pen, which was then brushed over with a chemical reagent to develop. This method wasn't as primitive as using lemon juice or other organic substances developed by heat, but it was still Secret Writing 101 because writing with a pen, usually a stylus, leaves deep impressions and scratches. It also leaves plenty of ink to analyze with tools like spectrometers. Another problem was that if an agent was caught, a bottle of invisible ink could be used as evidence against the agent in court.

As a result, along with the wet system came clever ways to conceal the invisible ink without having to bottle it unless it was disguised as perfume or medicine. The East German foreign intelligence unit reportedly concealed invisible ink in the molar of a tooth, just as the Nazis had done. A hollow tooth could also hide microdots and soft film. The agent could open the cover of the fake tooth by inserting a pin in a hole. Like the Germans before them also, the CIA concealed dehydrated heat-sensitive inks in many different ways, including disguising them as aspirin tablets.[8]

The situation at the CIA was even more primitive than that in the Eastern Bloc. CIA chemists recalled that "with secret writing, we were issuing systems that Caesar could have used during the Gallic Wars. We used systems that were developed during the First World War." As a result, SW chemists called themselves "lemon squeezers."[9] It is curious that the CIA seemed to have no historical memory of the extensive work done on secret inks during World War II by the Censorship Office's Technical Operations Division (TOD). Part of the problem was that the CIA's Technical Service Division (TSD) had former military medics on staff but no chemists and apparently no contact with other government offices. The personnel situation changed in 1951 when the chemist Dr. Sidney Gottlieb, later infamous for his involvement in the CIA LSD experiments, joined the CIA; he became head of the TSD's research and development program in 1959.

The Technical Service Division expanded from about fifty technicians in 1951 to several hundred scientists, engineers, artists, craftsmen, printers, and technical specialists a decade later. The East German Ministry for State Security expanded even more during the 1950s, from around fifty technicians to more than one thousand staff members in 1989 in a country a lot smaller than the United States. The KGB also had an enormous technical staff, even in proportion to the enormous country it worked in.[10] And both sides seemed to have expended about the same amount of human and technical resources on secret writing in the 1950s.

As far as methodological accomplishments go, the Eastern Bloc and Britain were the first to use the iconic Cold War system called the dry-transfer or carbon method. After the Cold

War a CIA chemist recalled that early methods were not much different from those used during World War II, when an agent used a wooden stick or toothpick wrapped in cotton to dip in a water-based ink. The agent had to steam the paper, write the message, resteam the paper, and then press it flat. Finally, the agent had to write a cover message on top of the invisible one.

When this anonymous CIA SW chemist was hired, his team "understood the Russians and the British did it a little bit differently and much more securely. My guess is that's when management finally realized we were far behind the curve in SW chemistry. . . . Why were we using this liquid stuff? Why couldn't we do it dry?"[11]

The dry-transfer system is best described as the secret-writing sandwich method: the spy places an invisible-ink-impregnated sheet between two normal sheets of paper (during World War II the German Abwehr wrote directly on the chemically impregnated sheet and the writing transferred to the invisible bottom sheet). The sandwich is then placed on a piece of glass or other hard surface. The spy writes on the top sheet and the invisible message transfers to the bottom sheet through the impregnated carbon.

During the 1950s the Eastern Bloc was already using the dry-transfer system more regularly than the West. Neither side seemed to have any historical memory of the way in which the Germans used a similar method to thwart iodine vapors during the world wars. The CIA did ask a private contractor to develop transfer systems during the 1950s.[12]

It wasn't until the 1960s that both sides embraced the so-called scientific-technical revolution and applied more sophisticated scientific methods to the creation and detection of secret-writing methods. Until then it had been relatively easy for both sides to develop enemy secret writing using the tried and true four methods of heat, silver nitrate, iodine vapors, and

UV lamps. The Eastern Bloc, in particular, caught many spies through interception and SW testing. After the wall went up in Berlin, the city of spies, it became increasingly difficult to catch spies, as rival espionage agencies developed more sophisticated methods. Such systems usually involved dry transfer. Since this method left few residual traces of scratches or imprints, and only a small amount of the substance was put on the paper, the ink was hard to find through chemical reagent methods. The dry-transfer method spurred on the new secret-writing arms race.

Probably the most innovative scientific breakthrough developed by both sides involved using only a tiny amount of chemical substance in the secret-writing message. This rendered the message almost undetectable. CIA chemists likened the process to "uniformly spreading a spoonful of sugar over an acre of land."[13] But the CIA never went public with the details of this magical substance.

We Play Spy Catchers

The CIA had even refused to release World War I secret-ink methods; they weren't about to declassify high-end Cold War systems and methods. Because of these restrictions in the United States, it was exciting to be able to reproduce a Cold War method involving such a minute amount of substance using methods and documents from Eastern Bloc archives. I had obtained a Stasi formula and method as part of the Stasi archives declassification process. By going in through the back door, I was able to open a window into top-secret writing methods also used by the CIA and other agencies.

Although there are hundreds of different chemical combinations that can be used to create SW, the trick is to find one that the other side can't detect with the four standard methods. Experimenting with the type of paper complicates the exercise, and the way the chemical substance is applied on the paper becomes part of the overarching method. Like the one-time pad in cryptography, one level of security is to keep changing the combinations. Another way is to find a substance that is hard to detect and hard to develop with the four standard methods. The East German Ministry for State Security SW scientists and the CIA developed such a method.

After I went home with the supersecret Stasi documents I had obtained in Berlin, I sought out chemistry colleagues to reproduce the Stasi method. My neighbor at the office, Dr. Ryan Sweeder, was enthusiastic about the project, and we quickly got to work on this intriguing experiment. We both wondered whether we had enough information to reproduce the secret method and to get the secret writing to develop, because the secret document did not include exact concentrations or measurements, but it did include the backbone chemicals used and some equations.

Cerium (III) oxalate played the starring role as the secret-writing substance in this method developed in the late 1970s. The beauty of the rare earth metal cerium as a secret-ink substance lies in its function as a catalyst. Since it speeds up the chemical reactions, only a small amount is needed on the secret-writing paper. Stasi scientists recommended spreading a small amount of the cerium—about five milligrams, or less than one five-thousandth of an ounce—in powder form on the sheet to impregnate the "carbon" used in between the top sheet, which the spy writes on, and the bottom sheet, where the invisible message is transferred.

The person receiving the message then brushes on a solution of manganese sulfate, hydrogen peroxide, ammonia carbonate, and a chelaplex, a chelating agent used to create a chemical compound in which metallic and nonmetallic atoms are combined.

And presto, the chemical reaction between the developer and the cerium oxalate creates a yellow, visible secret message—as it did in our experiment![14]

Get a Clue

The ball was now in the spy catchers' court. During the early years, the Stasi and other Eastern Bloc countries had been successful in catching spies who used secret writing. By the 1970s and 1980s chemical reagent methods for finding secret ink had become less successful as enemy agencies began to use this sandwich method with far less of the invisible substance on the paper. As a result, the Eastern Bloc started searching for physical ways to examine the paper. And this paid off.

Meanwhile, in the West, forensic scientists had developed the Electrostatic Detection Apparatus (ESDA) to search for indentations in questioned documents. Crime laboratories examined anonymous letters, pored over ransom notes, and analyzed extortion letters, or even seemingly ordinary business transactions, among other documents, for indentations that could penetrate three to four sheets of paper.

To develop indentations, the paper was placed on a bronze vacuum plate. The paper was then covered with Mylar, a transparent nonconducting polyester film. The paper was electrically charged and toner was poured on the Mylar, which developed the indentations. This forensic technique easily adapted itself to finding secret writing. In fact, by the late 1980s forensics scientists performed tests and discovered that they could reveal several dozens of types of secret ink containing salts, dilute acids, and most organic substances. Investigators felt that they had found the Holy Grail of secret writing: a universal developer.[15]

It seems that the Eastern Bloc Secret Service used its version of the ESDA more often to detect secret writing than did their

Western counterparts. Captain Renate Murk of the Stasi secret-writing technical division was an enthusiastic user of the Eastern Bloc's Nyom. One of the few women who worked at the Stasi's technical operations division, Murk was in charge of examining hundreds of envelopes and sheets of paper a year once they had been sent over from the mail censorship department. She followed the tried-and-true steps to determine whether the paper had signs of enemy secret writing. First, she examined the suspected sheet with the naked eye to find scratch marks or obvious disturbances in the paper. Then she used the slanted-light machine to see obvious indentations, and an ultraviolet lamp at different close proximities to detect luminescence. Various spectroscopes and chemical developers (from a list of almost a hundred reagents) ended the examination. When the Nyom arrived on the scene, testing with it became a standard part of the exam process leading to new successes in catching spies. Unlike chemical reagent methods, the physical Nyom method did not destroy the sheet of paper. It left no trace of investigation, so the message could be sent on to the recipient without revealing that it had been intercepted and tested.[16]

The New Contact Method

Of course, the secret-writing arms race was not limited to the KGB and its Eastern Bloc cohorts vs. the CIA. Sometimes the superpower method rubbed off onto smaller services, literally. In the mid-1980s, a British secret-writing technician at MI6 (foreign intelligence) developed a regular secret-writing message on the back of an envelope sent from an agent in Russia. That message appeared in a routine fashion, but another mysterious mirror-image Kiev address also began to develop on top of the main message.

British technicians concluded that when the agent had deposited the envelope in the mailbox, it had rubbed against another envelope with transferrable ink. Their minds raced, thinking about the possibilities of this transferrable ink. MI6 "mounted a systematic worldwide search for the magic pen" and sent every MI6 station secretary to local stationery stores to buy all available pens. If they found this magic pen, "it would be a superbly elegant, simple and deniable SW implement."[17]

Once they collected the pens, the technicians began testing all of them. They wrote a few words with each pen, pressed a piece of paper on top of the writing, then wiped it with developer. A few weeks later they identified the magic pen as a Pentel rollerball. They dubbed it the "offset" method, and MI6 officers used it routinely to write up notes after debriefing an agent.[18]

A similar method—dubbed the "contact ink method"—was developed by several other spy agencies and became as common as the carbon copy method in the 1980s. In fact, about half of all secret writers caught in East Germany used the contact ink method and the other half the carbon copy method.

West German intelligence developed a contact ink method, but it was more sophisticated than the British offset method. Case officers instructed agents to dissolve one teaspoon of cooking salt in about twenty teaspoons of boiled water. Into that solution they mixed fifty drops of sulfuric acid and forty drops of fountain pen ink. They were then instructed to write with a steel nib while wearing gloves to avoid fingerprints. Immediately after the writing was dry, they were to place the blank secret-ink letter on top of the written material with a heavy object for about two hours. It was then ready to send.[19]

Counterintelligence Mail Interception

Finding secret writing in letters sent through the post office is one of those proverbial needle-in-a-haystack jobs. Even so, post offices and censors have checked letters using chemical developers and infrared and ultraviolet lamps since World War I. During World War II, 14,462 United States censors opened a million pieces of mail per day from abroad. Of those, about forty-six hundred were passed on to the FBI because they looked suspicious, and four hundred contained important messages or used invisible ink.[20]

If democracies resorted to using censorship offices during wartime, the Soviet Union and Eastern Bloc totalitarian societies were notorious for opening and examining mail throughout the Cold War. This was no secret. Everyone assumed all letters were read.

It's not that an intelligence agency like the CIA did not *want* to intercept and read mail. On the contrary, U.S. agencies were eager to create a mail interception program. As a result they launched HTLingual in 1952, even though they knew it was illegal. The bottom line is that most intelligence agencies would like to have the free rein that the secret services enjoyed in totalitarian Eastern Bloc states, but American-style democracy, with its emphasis on civil liberties and its strong investigative media, makes it more difficult. (British, French, and West German societies were democratic, but media leaks were less common.)

Steam Street

With the license of a totalitarian regime and its all-encompassing state security apparatus, the Eastern Bloc developed highly refined mail interception methods. During the early years, mail opening and resealing was fairly primitive, done manually with very little use of technology. During the 1970s the Stasi introduced new technology to process bags and bags of mail more efficiently. Along with new departments and a stable of golden-handed engineers came a real technological marvel: the Automatic Mail Opener 10/10, which came into use by 1975. To complement the letter opener, an automated mail closer was developed as well.

In East Germany the Central Post Office worked closely with the Stasi's mail censorship department to filter mail. The sheer numbers were on their side. Although the MfS's central mail office in Berlin employed only 500 staff members, the fifteen regional offices brought the total number of staff members to 2,177. During the 1980s the Stasi examined a staggering ninety thousand pieces of mail every day before deciding which ones to scrutinize for secret writing. The new fully automated steam street conveyor belt system helped prescreen those thousands and thousands of pieces of mail more efficiently.

To start the steam street process, Stasi workers placed between three hundred and five hundred letters in a loading tray. From there, each letter was individually transported down the conveyor belt. A steamer inside the machine reached temperatures of more than two hundred degrees Fahrenheit to soften the glue on the envelope. Then a stream of air loosened the flap while warm air dried the letter. The fully automated 10/10 machine could open six hundred letters an hour as they rolled down "steam street." As engineers refined this method, they introduced an innovative method using cold steam to reduce the structural changes in the paper that occurred through heating.

After opening the mail, the Stasi used its newly developed letter-closing machine to glue the letters shut at the glue station. Then workers flattened the letter with a table press. The engineers boasted that their letter-closing rate per hour doubled with the machine: manually they could process only six hundred letters an hour; with the machine this rate jumped to twelve hundred.[21]

Mail Interception American Style: HTLingual/SGPointer

The CIA was well aware of the Eastern Bloc's effective and all-encompassing mail interception efforts early on during the Cold War. As a result the Agency had launched a "Probing" project as early as 1960. The Technical Service Division (TSD) sent hundreds of test letters in and out of the Soviet Union, recording for each the date and time of mailing, site of postal box, country of destination, type of letter or postcard, and whether it was written or typed, in order to find out whether the letters were opened and tested.[22]

The Eastern Bloc wasn't alone in filtering mail to search for enemy spy activity. Some years before the probing project, in 1952, the CIA's Office of Security launched a secret mail interception program at the request of the SR/Soviet Division. Initially code-named SGPointer, later HTLingual, the program was set up at the branch post office at LaGuardia Airport in New York City to intercept mail to and from the Soviet Union. Initially, representatives of the Office of Security copied only the exterior of letters, but later they opened envelopes and copied the contents of letters for closer inspection, which was clearly illegal.

Although there was a big uproar when the program was uncovered by the Church Senate Select Committee investigating the CIA's illegal activities in the 1970s, its efforts were modest compared with the Eastern Bloc's censorship and interception activities. Early on, program agents at the Manhattan Field Office examined 1,800 items a day and set aside 60 for closer examination (later estimates suggest the whole program netted some 215,000 intercepted letters), about one-fiftieth the number of letters the East Germans were examining. The CIA project never had more than a tiny fraction of the two thousand mail examiners East Germany employed.[23]

The program expanded a bit in 1961 when the CIA's Technical Service Division opened a laboratory in Manhattan. Agents were excited about the opportunity. Not only would the lab contribute to intelligence activities, but more important, it would offer "a workshop to test some of the equipment which the TSD has developed." The laboratory was charged with examining letters for "secret writing, micro/dots and possibly codes."[24]

Five years after the CIA launched the program, the FBI approached Postmaster General Arthur E. Sommerfield (1953–61) to set up a mail interception program. The Bureau did not know about the CIA's efforts, and the postmaster general got permission from the CIA to disclose it. After that, and because it was initially a counterintelligence program, the FBI received the lion's share of material from the mail interceptions.

Initially, the CIA counterintelligence staff developed a watch list with about three hundred to four hundred names on it. The FBI could also submit names. The early watch lists contained names of suspected Eastern Bloc citizens, including defectors and suspected illegals. But the list began to mushroom out of control, and by the 1970s it included names of senators and other government officials. CIA officials contemplated destroying such lists in case the press found out about, and exposed, the program: "In order to avoid possible accusation that the CIA engages in the monitoring of the mail of members of the U.S. Government, the C/CI may wish to . . . (a) purg[e] . . . such mail from the files . . . [and] cease the acquisition of such material."[25]

Although Postmaster General Sommerfield had been aware of the general nature of the CIA's program, he had approved copying only the covers of envelopes, not the contents. Initially, that is all the CIA did, but later agents surreptitiously opened the contents and then took them to the Manhattan office for closer inspection. Most subsequent postmasters general were not even apprised of the CIA's activities.

The CIA was very much aware of—some might say relished —the criminal nature of the program. In fact, agents come across as raccoons with face masks and striped clothes sneaking around New York City. They would stack up suspicious letters at the airport, slip them into their pockets, and bring them to the Manhattan lab at night. They steamed letters open and examined them between 5 and 9 PM, copied the contents, tested the letters for secret writing, resealed them, and returned them to the post office the next day. The CIA launched this program with "full knowledge" that a "flap" will put us "out of business" immediately and "may give rise to grave charges of criminal misuse of the mails by government agencies." They continuously used the word "surreptitious" to describe the program and acknowledged its illegality: "Essentially, there are two types of mail coverage: routine coverage is legal, while the second—covert coverage—is not." Later the FBI opposed any "covert coverage . . . because it is clearly illegal and it is likely that, if done, information would leak out . . . to the press and serious damage would be done to the intelligence community."[26]

In the 1970s the CIA nightmares came true. Details of the mail interception and examination leaked to the press, the story created a flap, the program was put out of business, and perpetrators were accused of criminal misconduct. The CIA, however, survived.

It is hard not to wonder what would have happened if the CIA's illegal activities had not been exposed. Would the CIA have become a rogue elephant breaking laws at will? (There was never any mention of legality in the Eastern Bloc's interception and analysis programs.) There is no doubt that it would have been advantageous to have the license to spy enjoyed by Eastern Bloc countries. It should be remembered, however, that even by 1973 the program was much smaller than its counterparts in the Eastern Bloc. Given the size of the United States, it was also much harder to screen materials.

While the CIA's major counterintelligence mail interception program was shut down in the 1970s, the Eastern Bloc's program was thriving. In fact, during the 1970s the Eastern Bloc scored its greatest counterintelligence successes. With increased use of technology in the 1970s, East German intelligence was able to catch thirteen CIA and West German spies between 1971 and 1982. Even without this technology, they had caught seven spies who used the post office in the 1960s.[27] The HTLingual program, in contrast, appears not to have led to successes in catching spies. We have only one reference to a Soviet illegal couple, "Mr. and Mrs. Robert Baltch," caught in 1963. They were let go because of a technicality: the FBI could not use the illegal wiretap and the mail intercept as evidence in court.[28]

Robert Glenn Thompson and His Microdot

Another spy who worked for the Soviet Union was not so lucky. American Airman First Class Robert Glenn Thompson was born in Detroit, but unlike the "Baltches'," his spy story began during the height of the Cold War in divided Berlin, not in America. Thompson was a high school dropout when he decided to join the U.S. Air Force at the age of seventeen in 1952. He served as a mechanic at various bases before he was assigned as a file clerk to the West Berlin Office of Special Investigation (OSI) at the Tempelhof Air Base in the American sector in 1955. He was cleared to handle secret documents, but nothing stamped "top secret."[29]

Initially, Thompson's main job at the OSI was to file three-by-five-inch cards, but he did not know their significance. When he found out, the new knowledge was a reason to hold a grudge

Robert Glenn Thompson

against the "Big Brother Is Watching You" state. He discovered that the OSI operated a joint mail interception program with the CIA to open mail in Berlin. All mail that came into Berlin from the United States was opened with a steam pencil slipped into the envelope; the contents were then photographed, and the names were placed on the OSI index cards by file clerk Robert Glenn Thompson. The air force had collected some sixty-eight thousand of these information cards.[30]

The first several years of Thompson's time in Berlin went well. He met and married a German woman named Evelyn and she gave birth to their first child while in Berlin. He was happy. But then one day in 1957 an air force buddy asked the six-foot-two, 250-pound, pockmarked Thompson to go out drinking at a bar with some friends. He did, but that bar episode ended in a court-martial for dereliction of duty when Thompson thought he lost his gun at the bar and replaced it with another one from the office the next day. He got a pay cut and was demoted a rank. His boss, a pudgy colonel with white hair and a mustache, found two guns in Thompson's pocket, because his drinking "buddy" had returned the one he took. Thompson's boss also sent his wife and child packing to the United States as further punishment.[31] Alone in Berlin and full of resentment, Thompson sought out female comfort at the numerous Berlin bars he visited.

Thompson's problems at work did not stop with the court-martial. In June 1957 Thompson went into the office on a Saturday to catch up on some work. Since it was the weekend, he hadn't shaved, and he loosened his tie to get more comfortable. It's not clear whether he was unlucky or whether this casual look had been a pattern, but the same colonel walked into the office and yelled at him for his slovenly appearance. This was the last straw for Thompson.

He stormed out of work and headed to his favorite watering hole, the Manhattan Bar in the funky neighborhood of Kreuzberg, at midday. His new girlfriend was there when he arrived. After drinking good-sized glasses of cognac, one after another, and venting to his girlfriend, he decided to do something drastic about his miserable situation: defect to East Berlin.

The wall hadn't yet been built, so he simply walked across to Friedrichstrasse and strolled down to Stalinallee, where he found a phone booth. He picked up the phone and dialed the East German secret police, using a number from his card files. The duty officer picked up the phone, talked to the drunken

Airman briefly, met him a few minutes later, and took him to a safe house nestled in the leafy green suburb in the Weissensee district, north of the Stasi's headquarters.

Initially, the German intelligence handlers considered Thompson a drunken loose cannon and rejected his bid for political asylum. Soon, though, they reeled him back in, outfitted him with a Minox camera, and had him photograph reams of secret documents about American intelligence operations in Berlin that were stored in the file room. After several months of copying secret documents, passing them on to the East Germans, and meeting in the safe house, Thompson received orders to transfer to a base in Montana. His handlers were more nervous about the impending routine move than Thompson was. Since he was moving to the United States, the East Germans handed him over to the Soviets. The KGB had more experience working in the United States and, during the late 1950s, more experience training agents in spy techniques. His new KGB handlers thought he needed training at their Black Sea spy school before starting his new mission in the States.

Thompson took a short leave—ostensibly for time in West Germany—and flew to a remote town on the Black Sea accompanied by his German handlers. The purpose of this retreat in an ornate eighteenth-century castle was to provide extensive training in secret writing, microdots, code, and radio messaging techniques. But the Soviets also wined him and dined him; they offered him women and vodka. They even had a going-away party for him. This was all in the name of building trust and friendship.

After flattering their new agent with attention usually reserved for VIPs, the Soviets were very serious about training Thompson in secure ways to communicate while he was in the States. The KGB's top-secret experts met with him and taught him the arcane arts of secret messaging.

Following breakfast during his first day at the spy castle, the Russians sent in a middle-aged woman to show him how to use the dry-transfer method of secret writing. She instructed him to use a desk with a hard glass top to write a message on. She told him to lay his writing paper horizontally, not vertically, and then described the sandwich method—placing the treated sheet between two regular pieces of paper. Write the message in block letters, she told him, then rotate the paper to vertical and write an ordinary visible letter. She gave him some impregnated carbons to stick in his address book for his trip home.[32]

When Thompson mastered the secret-writing lesson, the Soviets taught him how to use a shortwave radio and how to produce and use microdots. A man who walked with a cane and wore thick plastic black-rimmed glasses taught Thompson the arcane art of using and developing the microdot technique. By this time the Soviets had improved upon the Germans' World War II microdot techniques, the ones the FBI had dubbed masterpieces of espionage. Thompson became quite adept at learning the complicated steps of producing a microdot. He took a 35 mm reflex camera, clamped it on a chair back, made sure the document was lying flat, lined up the camera, and took the picture. Once the photos were developed, he put the negatives between two glass slides, took an ordinary piece of cellophane, and made it light-sensitive with chemicals.[33]

After that step, he placed his mikrat camera ("micro-pointer," he called it), a small brass tube about an inch and a half long with lenses on both ends, on a sheet of white paper on a flat surface. He held the glass plates with the negatives thirty-five inches above the mikrat and placed a magnifying glass above that. Then he directed the beam of a hundred-watt light through the magnifying glass, the negative, and the mikrat. When the negative was focused on a tiny spot on the white paper, he made an X, turned the light off, and slipped the cellophane under his mark. He turned the lights on, waited a few minutes, and developed the image with a regular commercial developer, producing a black dot on the cellophane, the size of a period at the end of a sentence.[34]

Even though this multistep microdot process sounds incredibly complicated, Thompson mastered it. He found the hardest part inserting and hiding the cellophane dot in a postcard. It took him days of practice to slit a postcard, insert the microdot in the right place, and reseal the envelope with flour and water to avoid black light.[35]

After his Soviet spy school training, Thompson flew back to Detroit and joined his wife. He did go to the base in Montana but decided he didn't want to work for the Soviets as a spy and never contacted them. This didn't deter them, and they sought him out in Montana. Trying to shake them loose again, he transferred to Labrador and obtained an honorable discharge from the air force, planning to set up his own business. By the early 1960s he made his way to Bay Shore, Long Island, and opened a heating oil business with his wife. The Soviets showed up yet again and convinced him to spy and report on people of interest.

Then one day Thompson noticed someone taking a picture of him and concluded that it was the FBI. He never attempted to run; instead, he waited for agents to show up on his doorstep. The FBI took its time, keeping him under surveillance for many months, and when agents arrived, it wasn't to jail him, but only to question him. And question they did. He told them everything, giving them the evidence they needed to arrest him. Thompson recalls twenty meetings with the FBI before his arrest. He thought that his cooperation would earn him a reduced sentence. Instead, he was eventually sentenced to thirty years' imprisonment. After spending more than ten years in jail, he was traded in a spy swap in 1978 and went to East Berlin under the wing of the East German intelligence service. He lived under the alias Gregor Best and worked for foreign intelligence in false flag operations, pretending to be a CIA recruiter.[36]

The Thompson case was the first time the American public had the opportunity to learn about real Soviet secret communication techniques. When Thompson told his story to the *Saturday Evening Post* and the magazine published a riveting two-part series in June 1965, including photographs of the microdot technique, it was the height of spy fever in America. By that time, three of the wildly popular James Bond movies had been released, including *From Russia with Love*.

This wasn't the first time secret writing and microdots had been synonymous with nefarious spies, nor was it to be the last. But like the German spies in World War II America, the FBI was excited to learn about the Eastern Bloc's latest techniques and methods, often at the expense of evaluating the information transmitted. In this case, using secret writing did not lead to Thompson's arrest, though it did constitute evidence of spy activity. It turned out that Dimitri Polyakov, a Soviet GRU deputy resident who volunteered to spy for the FBI, had tipped off the Bureau to a clumsy Russian third secretary at the United Nations, Fedor Kudashkin, who had sought out and worked with Thompson. Kudashkin was quickly expelled.[37]

The KGB and other Eastern Bloc security services were not the only spy agencies that used the microdot during the Cold War. The CIA was also quite clever in concealing its microdots. During the early 1960s it hid a microdot in a riveting scholarly journal called *History and Significance of Earth Measurement*. Who would have thought that a microdot was hidden in the first letter "a" of the word *Invardrähte*? The indefatigable technicians at the East German secret police found the microdot in a routine examination of a number of brochures. After eyeballing the page of the journal and looking at it more closely with a microscope, the technicians saw a rectangular dark framing around the pertinent letter near the bottom of the page. Further examination showed disturbances in the paper fibers around the letter. When they magnified the microdot with a 140× powered lens, they found a message typed in capitals in the style of a telegram, directing an agent to a meeting abroad at an equestrian statue. The message also provided detailed instructions for surveillance detection: the agent was to stop under a train station clock for ten minutes, then leave for some window shopping in the town before meeting the handler at the statue. If the agent thought he was under surveillance, he was supposed to take a newspaper out of a bag and hold it in his hand until he got to the statute. That would be the sign to abort the meeting and delay the meet until the next opportunity.[38]

Even though microdots were hard to detect, they clearly weren't foolproof. Since they were also hard to use and view, the CIA had reservations about providing them to agents. But when it was necessary to use microdots, it was also necessary to provide the receiving agent with a small viewer that was easy to hide. The Agency developed many concealable microdot viewers (like Reif's viewer, concealed in a felt-tip pen). The smallest one was a modified Stanhope lens, also known as a "bullet" lens, which could fit in the corner of an eyeball, a cigarette, or a bottle of ink. If you take a penny and lay the lens on it, it is about the length of Lincoln's head. The agent would paste the microdot on one end of the lens and view it through the other end. It magnified the image thirty times.

During the 1950s, the CIA ordered a hundred of these lenses from a novelty store only to discover that each one came preloaded with sexy pinups. In order to discourage gawking and facilitate viewing of the microdot, the CIA had to remove the sexual images. Needless to say, this lens wasn't popular with agents because it was so small, but it did score several operational successes, including the reading of a microdot hidden in the tiny dried fish so popular in Asia. This Stanhope lens was allegedly as big as a grain of rice and fit under a Band-Aid on a foot sore; one wonders how big the image became with such a small lens![39]

Microdot Imaginations

It's hard not to be fascinated with the microdot: it's tiny, it holds secret information, and it's associated with intrigue. Therefore it is not surprising that popular culture latched onto the microdot during the Cold War. If the reader knows what a microdot looks like, it is probably from having seen one of two successful Cold War spy thrillers that appeared shortly after Thompson was sentenced and his case received so much publicity. The movie *Arabesque*, released in 1966 and starring Gregory Peck and Sophia Loren, is a romantic thriller about a professor of hieroglyphics (Peck) who is hired by an Arab oil magnate mogul, Beshraavi, to decipher a secret message. The political fate of the Arab nation hangs on this undeciphered cipher. At first, the professor struggles with the hieroglyphics because they seem to convey trivial nonsense. Then one rainy day, when he is traveling in a car with Yasmin (Loren), he places the hieroglyphics message on the wet windshield. As the message gets wet, the ink dissolves in the rain, but the pictographic eye of a goose pops out and glistens. The professor quickly finds a microscope. When he reads the message about an assassination plot against the prime minister, the chase is on, as the professor and Yasmin attempt to warn the target.[40]

A year after *Arabesque* appeared, the James Bond spy film *You Only Live Twice* was released. Tiger, Bond's Japanese ally, informs Bond that his team has found a microdot on a purloined

SPECTRE photograph of the cargo ship *Ning-Po*. When Tiger en-
larges the microdot, the message states that the tourist who took
the photograph was killed for security reasons.[41]

Visible Spies

The Eastern Bloc seems to have caught more spies using secret
communication methods than their Western counterparts. Reif
was just one of several dozen spies who were caught because of
East Germany's tightly woven net. The mail department took a
leading role in catching Reif and worked closely with counterintelligence; the Sector for Operational Technology (OTS) offered technical assistance. Of course, blanket surveillance and
unrestrained interception of mail made it easier for Eastern Bloc
counterintelligence to find spies communicating with invisible
ink. But intercepting letters was only the first step. If a message
could not be made visible, a suspect could not be convicted as
a criminal. In the seventies, both sides improved their rate of
success in making invisible messages visible. By the end of the
eighties there was a precipitous drop in the use of secret writing,
as agents started to communicate via programmable calculators,
like the Casio, or computers.

The CIA and West German intelligence often caught spies
because of defector information, not mail interception, even
though evidence from spy paraphernalia like cameras and secret-writing devices was used in court. By contrast, the Eastern
Bloc and KGB seemed to have been fairly successful in catching spies by intercepting and deciphering secret writing. The efforts of the mail interception department, which had co-opted
the post office, certainly played a large role in prescreening suspects. The number of pieces of mail and suspected material examined lays bare an operation of enormous intensity and effort.
The technical division also poured a large amount of money and

resources—including the vast human resources of the men and women scrutinizing documents—into creating and detecting secret writing.

During the Cold War both sides used similar secret communication techniques, but democratic oversight limited the West in carrying out the successful first step of intercepting possible secret-writing letters. This does not mean that we should approve of illegal activities by our intelligence agencies. It might mean that democratic societies need to rely more on diplomatic means to achieve their goals; losing the spy wars may be the price of winning the larger ideological battle.

Prisoners

Of course it wasn't just spies who used secret writing during the period of the Cold War. But they were the ones who had become pawns in the spy games of international intrigue. They were the ones who accelerated the arms race between the invisible-ink creators and detectors producing more sophisticated chemistry and methods. And, paradoxically, they were the ones who became most visible after they were caught.

Less visible were prisoners and lovers who sought to communicate privately. During the period of the Cold War, prisoners behind domestic bars, as well as prisoners of war abroad, devised ingenious methods to pass on secret messages to the outside world. One of the most striking stories of secret communication by a prisoner of war is that of James Bond Stockdale (1923–2005), a navy pilot shot down by the North Vietnamese on September 9, 1965.

After Stockdale ejected from his fighter plane, he parachuted into a small village, where he was beaten before being taken as a prisoner to the Hoa Lo prison—known to American prisoners as the notorious Hanoi Hilton—for the next seven and a half years. While he was there he was one of the main organizers of a code of conduct for prisoners that included how to deal with torture and how to communicate secretly.[42]

At first Stockdale started communicating in code with his wife, Sybil, in order to transmit a list of fellow prisoners. He simply referred to several football buddies from his class at the U.S. Naval Academy, even though they had never played football with him. Sybil told U.S. Naval Intelligence about this light code, and her navy contacts enlisted her help in passing on a secret message. The idea was to send a photograph every month in the one letter he was allowed to receive. Once a routine was established, officers would give her a specially prepared photograph. She initiated the communication with a photograph of a fake mother-in-law.

Shortly before Christmas of 1966 the North Vietnamese gave Stockdale several letters at a filmed propaganda session; one of them included the photo of his "mother-in-law." When he was back in the privacy of his prison cell, he was puzzled by the photograph and decided to destroy it, but then thought: "But wait a minute—it's dumb to throw away something from the States without doing more with it. James Bond would soak it in piss and see if a message came out of it." He filled his drinking water jug with urine and soaked the image. At first nothing happened. Then, after half an hour, he took out the photo and laid it on his bed slab to dry. Gradually, he started to see a speck, then a small print paragraph appeared on the back of the picture. He read it over and over again to burn it into his brain before his jailers picked up the pail of excrement into which he planned to rip the photograph. Twenty years later he was able to reproduce the remarkable message in his memoirs:

The letter in the envelope with this picture is written on invisible carbon. . . . All future letters bearing an odd date will be on invisible carbon. . . . Use after you write a letter. . . . Put your letter on hard surface, carbon on top and copy paper on top of that. . . . Begin each carboned letter with "Darling" and end with "Your adoring husband." . . . Be careful; being caught using carbon could lead to espionage charges. . . . Soak any picture with a rose on it. . . . Hang on.[43]

Even though Stockdale knew the danger of an espionage charge, he continued to write secret letters with information about the prison camp, identities of prisoners, and the torture methods his captors used—"hand and leg irons—16 hours a day." When the CIA learned of these secret-photograph letters, it stepped in and helped devise another secret-communication link.[44]

Domestic prisoners also need to communicate, but rarely do they have access to the secret-communication methods of a navy officer like Stockdale. Instead, they need to rely on their wits and organic methods available in prison: semen, urine, saliva, lemons, oranges, and diluted blood. By the 1980s the notorious prison gang the Aryan Brotherhood had developed ingenious methods for communicating. Aside from employing Francis Bacon's binary code, they used urine or citrus juice as a writing substance to orchestrate carnage seventeen hundred miles from their Rocky Mountain supermax prison. In August 1997, T. D. "The Hulk" Bingham, the "walrus-mustached warlord," sent out an order written in invisible ink to his confederates at the Pennsylvania Federal Prison to charge black inmates with shivs. Brotherhood prisoners killed two of them. The secret message had read: "War with DC blacks, T.D."[45]

Lovers

Lovers preserved their privacy and intimate thoughts by communicating with relatively primitive forms of invisible ink. Like a sibling who writes a diary in invisible ink to prevent a family member from prying, they didn't need sophisticated inks or technologies to protect their intimate thoughts. Thousands of lovers throughout history have communicated romantic thoughts this way, and most of their stories have remained hidden and secret to the rest of the world.

For centuries, adulterous lovers followed Ovid's second-century poetic advice to dupe their partners by writing secret messages with milk. By the seventeenth century, lemon juice and alum had become more popular for lovers trying to conceal their intrigues. Even for faithful lovers, these primitive inks provided a way to communicate privately and romantically. Writing invisibly also helped men and women who felt uncomfortable expressing their intimate thoughts in person.

By the eighteenth century, when cobalt chloride fire screens became the rage in romantic Paris, special sympathetic inks were developed especially for ladies ("Encre pour les Dames"). In crystal form cobalt chloride sparkles an intense red, it turns blue when heated, and when it is dissolved in water, the solution is pink. In the twentieth century, this romantic ink became available through mail order; pulp magazines advertised bottles of romantic ink: "Write invisible love messages in passionate-red invisible ink which only you and your lover can make appear and disappear. Protect your love life."[46]

During the Cold War the physical barrier of the Berlin Wall cut off families, friends, and lovers. Many separated people communicated in code, but lovers preferred the romance and privacy of invisible ink. The wall made hearts grow fonder. The notion of lovers separated by a wall is not a modern invention. In the myth of Pyramus and Thisbe, the lovers' parents oppose their liaison and build a wall between the families' houses. Just as a wall separated the star-crossed mythic lovers, so too did the Iron Curtain separate countless lovers.

When Mano Alexandra was a young woman living in Prague, she fell in love on the Isle of Kampa (also known as the Venice of Prague), located on the Vitava River in the middle of the magical city. With its Disneyesque Old World setting, cobbled winding streets dotted with intimate cafés and romantic restaurants, Prague was the scene of her first major love affair. But then the worst possible scenario happened. The two lovers were separated by politics: in 1968 the Soviets invaded Czechoslovakia, and Alexandra escaped the country and left her lover behind. The separation was painful. They both so wanted to see each other, to fall into each other's arms and make love. But they couldn't see each other in the flesh, so they kept their love alive by writing to each other every day. They knew that government agents read their correspondence, so they could not disclose their deepest desires. How Alexandra longed for lovers' invisible ink.[47]

Years later, when it was too late, Alexandra discovered the French cobalt chloride invisible ink that other lovers had used since the eighteenth century. When she experimented with the ink and saw the invisible words appear blue, like magic, she was enchanted and reminded of another Prague love story, one her grandmother used to tell her that took place shortly after World War I. In that tale, the star-crossed lovers used invisible ink to communicate their deepest love and desires while keeping it secret from the woman's aunt, who was passing on the letters.[48]

Scores of other separated lovers developed creative ways to pass on messages secretly during the Cold War. When an American artist and her Yugoslavian lover were apart, they communicated by writing messages in black ink, then covered them with watercolor paintings. The lover receiving the painting washed off the watercolor to read the message.[49] Although they didn't know it, this was a form of steganography (hidden writing), the subject of the next chapter.

Hiding in Porn Sites

13

USA TODAY STARTLED THE world in February 2001 when Jack Kelley reported that terrorists might be hiding blueprints to attack the United States of America in X-rated pictures on pornographic websites. The very thought was enough to galvanize federal law enforcement officials into action. But unlike e-mails and phone calls, tracking messages in digital photographs posed a special challenge, because both the message and the very act of communication are hidden. Investigators never did find those nefarious messages in X-rated pictures, at least not in 2001. For the next ten years they searched and searched, to no avail. Niels Provos and Peter Honeyman, leading information security specialists at the University of Michigan, reported that they "analyzed two million images" from the eBay auction website but did not "find a single hidden message."[1] There was nothing there, or so they thought.

Ten years later, computer forensics experts from the German Federal Criminal Police (BKA) hit pay dirt. In May 2011, Maqsood Lodin, an Austrian national on a watch list, was detained, questioned, and searched by Berlin police because he had just returned from Pakistan and was suspected of having attached himself to militant Jihadists on the Pakistan-Afghanistan border. In the course of the search, police were surprised to find digital storage devices and a memory card hidden in the terror suspect's underwear. After receiving the material, computer forensics experts spent weeks trying to break in to a password-protected folder. To their great astonishment investigators found a pornographic video called Kick Ass and a file labeled "Sexy Tanja." Embedded in those porn films lay 141 text files containing documents describing al-Qaeda operations and plans for the future, with titles like "Future Works," "Lessons Learned," and "Report on Operations." Most chilling were plans to seize hostages on passenger ships, dress them up in orange prisoner jumpsuits, and execute them in front of rolling cameras.[2]

Terrorists hid messages in plain sight, secreting unencrypted files within the porn videos via what professionals call image steganography, or "stego." As we have seen, the word "steganography"—hidden writing—is a coinage by the fifteenth-century abbot Johannes Trithemius. And even though the ancient Greeks didn't use the Greek word (*steganos* means cover or roof), they practiced the art of hidden writing, using wax tablets and shaved heads to hide secret messages. Thousands of years later, the digital age reincarnated the ancient art and turned it into a sophisticated science.

And as hidden writing became digitized, it began to catch up with, if not surpass, cryptography in sophistication and importance. Computer power, the Internet, and the fear of terrorist attack accelerated its development. In fact, digitization and governmental limitations on the use of encryption have propelled stego to the forefront of secret communication security.

The most common digital form of steganography conceals messages in audio, image, or video files. The easiest and least secure way to hide a message is to open a JPEG file with a text editor and append the text at the end of the content. Of course, the file would be larger and the hidden text could be found very easily. As a result, computer experts have developed sophisticated software tools that manipulate the bits in the media file to camouflage the secret text. The idea is to make sure the image looks the same to the human eye, and the song sounds the same to the human ear.

Since every digital picture is made up of thousands of minuscule screen dots, the naked eye won't notice if the sender replaces an image dot with a secret message. The computer identifies every one of these dots with a long string of binary numbers—zeroes and ones. Apparently a few of these numbers don't matter much, so they are called the "least significant bit" or the most redundant bits of data. They are invisible to the human eye.

In an audio file the sender can use a sound bit that humans can't hear, while with an image, he or she can remove the least significant bit of color. Stego hides its data in those little image or audio bits. Stego makers can buy or create software programs that use an algorithm to embed the data in an image or sound file.[3] The receiver then retrieves the secret message with a password, just as the BKA agents did with the porn files.

Stego professionals can transform innocent-looking image and music files into vehicles for the binary code, which includes 0s and 1s in certain sequences, like 1000011. The numbers code for letters or images—A for 1000011, for instance. Stego pros have developed embedding methods and substitution or insertion-based ways to hide data. They often use the least significant bit because it is the last bit, or the number 1 at the end of the sequence above.[4]

Stego Makers/Stego Breakers

The widespread use of the Internet has brought with it an increase in cybercrime, which has in turn led to the evolution of a new type of detective: the computer forensics investigator. Since 2001 there has been a marked increase in computer forensics experts who specialize in digital steganography. They have professional organizations and conferences, and the once obscure field has become better known. These investigators often work in the shadows. Many are reluctant to reveal how, whether, or where they detect stego on the web because it might harm national security.

Neil F. Johnson, a steganography consultant, lecturer, and president of his own company in the Washington, D.C., area, thinks "it's foolish to disclose what I'm scanning for, whether I'm scanning and whether I'm detecting anything" because it would tip "one's hand." He has stopped publishing his research on how to detect steganography because criminals or terrorists could devise countermeasures to thwart him. He also prefers to remain personally anonymous. His face is obscured and he is wearing dark glasses on his website; there is no contact information.[5]

Some stego breakers are more forthcoming. In 2001, Chet Hosmer, whose company Wetstone Technologies in New York works under contract with the U.S. Air Force, found that about "0.6 percent of millions of pictures on auction and pornographic sites had hidden messages."[6] Even though Niels Provos came up with nothing when he searched the eBay auction site, he still thinks the technology should be shared openly. He published his research because he believed in the free exchange of ideas, even with changed political context after the 2001 terrorist attacks. Provos's program used a statistical analysis of downloaded images to detect hidden messages.[7]

Whatever stego breakers' views on sharing sensitive knowledge, it is clear that a new arms race has developed between the stego makers and the stego breakers. While in the early 2000s only several hundred stego programs existed, by 2012 more than one thousand programs existed for purchase or free download.

The Prisoner's Problem Digitized

Another way to look at modern stego is to digitize the so-called prisoner's problem. When John Gerard miraculously escaped from the Tower of London, he communicated with the outside world in primitive invisible ink. Of all the people profiled in this book, prisoners have had the hardest time communicating secretly. Their letters might be censored, and the prison warden might not pass them on if they are suspicious or written in code. Gerard communicated through the warden by gaining his trust and by handing him rosaries wrapped in paper impregnated with secret ink to pass on. In this way, he crafted an escape plan with his confederates on the outside without raising the suspicion of the prison guard. In a sense, the prisoner's problem is everyone's problem. There is always the threat of enemy interception when people try to communicate secretly.

In the modern world of cyberspace and the Internet, the prisoner's problem can be overcome by communicating invisibly using stego. In 1984, Gustavus Simmons, a cryptographer at Sandia National Laboratories, proposed a solution to the prisoner's problem using a "subliminal channel" in cryptography. With the advent of digital steganography in the 1990s, a logical step was to apply Simmons's scenario to images instead of ciphers. In Simmons's model, two prisoners convicted of the same crime are placed in separate cells far away from each other. But they desperately need to communicate with each other in order to craft an escape plan. The only way they can communicate is through the warden, who will throw them into solitary confinement if she catches them. A solution to the prisoner's problem is to hide the information in innocent-looking pictures exchanged with the other prisoner.

The prisoner's problem and solution is a nice model to use when thinking about other situations where two people need to communicate secretly but face the threat of interception. The prisoner's three-part framework is ever-present when communicating using steganography: the cover object (the image); the stego object (the secret message); and the stego key (a password or a place to look for the hidden message).[8] The stego key is another digital transformation of a method used in the spy world to communicate: the dead drop. During the Cold War, spies used a dead drop to deposit tools of the trade or secret messages anonymously. It was a form of impersonal communication that often occurred in remote places like cemeteries or sometimes under city footbridges and drainpipes. In the digital world, the dead drop can be anyplace on the Internet, such as an eBay auction or a pornographic website.

Unlike the prisoner scenario, terrorists or spies communicating through an auction website like eBay or an anonymous pornographic website don't have to deal with a middleman or -woman. That is the beauty of the digital dead drop sites: the sender of the message can simply post a hidden message in an image of an auction item and the receiver simply needs to download the item and retrieve the message like the thousands of customers conducting transactions every day.

Russian Spies

This is exactly what a ring of ten Russian agents based in New York City, Boston, and New Jersey did for many years before they were arrested in the summer of 2010. The spies replaced the old-fashioned physical dead drop with the up-to-date digital dead drop. The "illegals"—spies who have adopted someone else's identity—communicated like the rest of the world: they used the Internet. Unlike the rest of the world, however, they were outfitted by their spymasters with sophisticated stego software not available commercially. Moscow Center instructed them to communicate through images on ordinary websites. Their main mission was to "search and develop ties in policymaking circles in US and send intels [intelligence reports] to C [Moscow Center]."[9]

The ringleader of the group was a "Richard Murphy," who

lived an ultrasuburban lifestyle with his "wife," Cynthia, and two children in ultrasuburban Montclair, New Jersey. A neighbor commented that the "Murphys" must have taken a course in "Suburbia 101." In between grilling hamburgers on the backyard barbecue while sipping Bud Light beers and eating brownies baked in the shape of the statue of liberty, they were frequent users of digital dead drops. When the neighbors weren't watching, Richard and Cynthia Murphy booted up their laptop computer, visited public websites, and downloaded innocent-looking pictures.[10] One day in April 2009, the Murphys clicked on the intelligence agency–grade stego software icon from Moscow Central and the image began to reveal the text with its secret electronic message:

> C plans to conduct flash meeting w. A [Murphy] to pass him $300K from our experienced field station rep (R) [a Russian government official]. Half of it is for you. Another half is to be passed to young colleague (known to you) in fall '09–winter '10.

The message proposed a meeting at the suburban North White Plains train station. It was "quiet and deserted on weekends. No surveillance cameras."[11]

Although there were no public surveillance cameras at the station, Moscow Center and the spy couple never imagined that the FBI would conduct its own video surveillance ("pursuant to judicial order") when the "brush pass" of the money took place in the summer of 2009. The exchange of money happened just as Moscow Center advised: the Russian government official silently placed a Barnes and Noble shopping bag with the money in Richard's Murphy's backpack as he was walking up the stairs. They continued to walk once the brush pass was completed.

This wasn't the only exchange the FBI witnessed. It had been spying on the couple, and the rest of the ring, for some time: it tapped their phones, read their e-mails, placed listening devices in their homes, and installed video cameras in public places and hotels. In July 2005 investigators searched the Murphy's home. While there, they hit the jackpot. They found and photographed a piece of paper with a twenty-seven-character password on it. When they returned to the office, the FBI agents tried out the sequence outlined on the piece of paper. Pressing the Alt, Ctrl, and e keys and entering the password, they found they could access the steganography software program stored on the Murphys' password-protected disks.[12]

The FBI also copied and analyzed the hard drive of the Murphys' computer and found links to website addresses they had accessed. The agents then visited some of the sites and downloaded images from them. When they analyzed the images with the Russian steganography program, they found more than one hundred readable text files, including the one about meeting at the surveillance camera–free suburban train station.[13]

Even though it appears that Moscow Center has updated its spy technology to fit in to contemporary society, Russian agents are still capable of spying the old-fashioned way. When two other members of the spy ring based in Yonkers, New York, were planning a spy trip to South America, the FBI eavesdropped on the following conversation between Juan Lazaro and Vicky Pelaez:

> LAZARO: When you go [to the South American Country] . . . I am going to write in 'invisible' . . . and you're going to pass them all of that in a book . . .
>
> PELAEZ: Oh, o.k. . . .
>
> LAZARO: I'm going to give you some blank pieces of paper and it will be there . . . about everything I've done.[14]

Lazaro planned to use invisible ink to provide Pelaez with information for their Russian handlers in South America. Even with modern technology, it seems that invisible ink is here to stay forever.

One of the most memorable Russian spies, Anna Chapman, used yet another mode of communication to send messages to her Russian handlers from a Starbucks at 47th Street and Eighth Avenue in New York. Red-haired and sexy, Chapman became the darling of the media and the heartthrob of many men when her photograph started circulating in the media. She became known as the "red under the bed," to suggest a seductive femme fatale, but she was, in fact, a well-trained, tech savvy spy who also used her brain. She communicated with her Russian handler via wifi on her laptop computer.[15]

This is how the FBI described her activities at the Manhattan Starbucks: "CHAPMAN was seated near the window of the Coffee Shop and had with her a bag (the 'Tote Bag'). After approximately ten minutes, I observed a minivan pass by the window of the Coffee Shop." The FBI agent goes on to describe the surveillance, including the detection of wireless communication, using commercially available technology, between Chapman and the Russian office in the van.[16]

DNA Microdots

Steganography in the early twenty-first century wasn't limited to the Internet or to terrorists and spies. It became a multimillion-dollar industry, with buyers ranging from corporations fighting industrial espionage to government agencies battling terrorists and spies to hackers breaking into networks. The new Gold Bug has bitten natural scientists as well.

In 1999 a research group at Mount Sinai School of Medicine, headed by Professor Carter Bancroft, came up with the idea of hiding messages in DNA microdots. In the course of his work as a molecular endocrinologist and biophysicist, Bancroft became interested in using DNA as "a kind of organic computer" and that got him "thinking about ways to use DNA outside the bounds of a living cell."[17]

Inspired by the legendary microdot method used by spies during the Second World War, the researchers took the "enemy's masterpiece of espionage" "one step further" to hide the message not just once, like the Germans, but twice, so it is "doubly steganographic." First they "camouflaged" a DNA-encoded message within the enormous and complex human genome, then they concealed it again by transferring the DNA onto a dot and pasting it onto a period in a letter sent to the recipient, thus creating a microdot.[18]

Because DNA is made up of building blocks or nucleotides—designated A, C, G, and T—it can be rearranged millions of ways. The researchers first encoded each letter of the alphabet, punctuation marks, and numbers with three nucleotides. For example, the letter A is encoded with CGA. Then they created their sixty-nine-nucleotide-long DNA stranded message using the key to re-create the World War II message "June 6 invasion: Normandy." They marked each strand with "primer" sequences of twenty nucleotides at the beginning and end, to further deceive a decoder. Then they pipetted the DNA message onto a period on a piece of filter paper, cut the period out, and placed it on a real period in the letter they sent with the secret message.

It's hard to imagine a spy using the technique, but molecular biologists engaged in secret work might find it useful to hide their research from competitors. In addition to cryptographic uses, the researchers thought the new method had potential for corporate security, and especially, for "tagging items of interest" with a DNA watermark.[19]

Even if this biological double stego didn't catch on with

spies, it did inspire the scientific world. The Defense Advanced Research Projects Agency (DARPA) challenged scientists to come up with chemical instead of electrical methods of communicating (we can only imagine how the kooky-project-creating agency wants to apply this form of communication). The well-known biologist Craig Venter wrote his name and several quotations onto the DNA of partially synthetic bacteria. A new field of biocryptography and steganography emerged by 2009, along with novel biostego methods.[20]

Another group of chemists funded by DARPA's chemical communication program used bacteria for message hiding and called it steganography by printed arrays of microbes (with the unfortunate acronym SPAM). This technique appears simpler than the DNA microdots because the SPAM-encoded messages light up in the colors of glowing bacteria and can be seen with light-emitting diodes or an iPhone. The secret message can be sent through the post office, unlocked with antibiotics, and deciphered easily. A sender using the right E. coli strain could add a time stamp to the message or let it self-delete, just as the recordings on *Mission Impossible* self-destructed.[21]

Even if bacteria are simple organisms and all the receiver needs is an iPhone to read the message, a spy would still need to pack a minilab when on a mission to send a message. Like the DNA microdot, however, SPAM could have practical applications in watermarking genetically modified organisms with "biological barcodes" or in preventing counterfeiting, according to lead scientist Manuel Palacios. But of more interest to DARPA, the scientists argue, it could "enable biometrics" or "communication through compromised channels."[22]

By the early twenty-first century, the ancient art of hidden writing was reborn in a hi-tech world. No longer confined to warfare, hidden writing appealed to users as diverse as terrorists hiding messages in pornographic websites and scientists looking to combine information hiding with their own fields of molecular biology or chemistry. The spies, of course, will be here forever.

As to the historical evolution of hidden writing, it is clear that the little sister steganography has finally caught up with her big brother cryptography in sophistication and complexity. Digital images and life-encoded messages have taken the place of wax tablets and microdots. Even if those messages aren't always hiding in porn sites, they *are* always hiding in plain sight.

Epilogue

IN APRIL 2011, THE CIA sent out a press release with great fanfare; it had finally released invisible-ink recipes dating back to the First World War. It even posted the six documents, totaling sixteen pages (with many duplicates), on the National Archives electronic reading room website for the whole world to see. This certainly was a momentous event. Since 1998 Mark S. Zaid, a Washington, D.C., lawyer, had been filing lawsuit after lawsuit, based on the Freedom of Information Act, on behalf of the James Madison Project to reduce secrecy in government. The recipes were the National Archives' oldest classified documents.

The CIA blocked release of these documents on national security grounds. The Agency claimed that the papers could provide "building blocks" for unlocking the CIA's modernized methods of producing or detecting secret writing, that they could aid in identifying current "vulnerabilities," and that CIA agents could still use the invisible ink. In 2002, a federal court ruled in favor of the CIA. Nine years later, CIA Director Leon Panetta deemed the documents no longer sensitive because "recent advancements in technology made it possible to release them."[1]

When I examined these formerly classified documents, it wasn't only a letdown but also a shock. First, most of the information in the files had already been published, or I had found it in other archives or collections. Second, the files did not even originate with the CIA. They belonged to the Office of Naval Intelligence (ONI), but somehow had made their way to the CIA. This shouldn't be a surprise, as the CIA was not founded until 1947, and its predecessor, the OSS, was founded in 1942. Finally, because of mandatory release of files relating to Nazi Germany, the CIA had been forced to declassify OSS records from World War II that were much more sensitive than the harmless World War I release. In fact, I used these rich little-known World War II sources for many chapters in this book. Furthermore, there are other, much more important secret-ink files from World War I originating with military intelligence. Some have recently been released to other government agencies, and some have disappeared.

As my manuscript was going to the publisher, I found another set of World War I invisible-ink documents released by the National Security Agency, but the agency, living up to its nickname "No Such Agency," didn't send out a press release. The NSA release is much more significant than the CIA's release. It boasts 895 secret-ink samples, tests, and effects of developers, a much more thorough and complete set of documents than the CIA file release. Apparently, the World War I secret-ink files were distributed to various government agencies after the war and passed down to successor organizations.[2]

If anything, the 2011 release should be an embarrassment to the CIA. One set of documents is about German secret-ink formulas dating from June 14, 1918. Apparently French intelligence sent them to the ONI, which translated them. The one-page document includes formulas for invisible ink made from a mixture of pyramidon and aspirin. It also includes a two-step development process.

As we have seen, during World War II, toothpick-wielding Nazis made headline news with their spy novel–like techniques.

They dipped toothpicks in pyramidon and passed secret messages about America's military strength to German military intelligence. This public knowledge is available in books and in newspapers like the *New York Times*. No secrets here.

What one can learn is that the more things change the more they stay the same with invisible ink. The Germans had the audacity to use the same secret ink during both world wars.

The second set of documents offers a secret-writing method in which the agent impregnates a handkerchief, starched collar, or starched shirt with nitrate of soda and starch in water. After the items are soaked and ironed, they look like starched garments. To use the ink the agent simply dips the items in water and squeezes them. It wasn't just toothpick-wielding Nazis who made headline news. During World War I the Mata Hari–like spy in America, Madame Maria de Victorica, used her silk scarves as weapons since they were soaked in invisible ink that could be squeezed out after the scarves were soaked in water. This ink can be developed with iodite of potassium, sometimes considered a universal developer. The starch-soaked shirts in the CIA's stash are nothing new either.

Another document experiments—"not successful"—with a way to open sealed letters without detection. And finally, another document from the Department of Commerce lists several other invisible-ink samples, including substances discussed in this book, like cobalt chloride.

The final document is perhaps the most embarrassing. It includes an article written by Theodore Kytka, a San Francisco handwriting expert, called "Invisible Photography and Writing, Sympathetic Ink, etc." Even though this copy of the article is stamped confidential, not for publication, it was published in the *American Journal of Police Science* in 1930. I used Kytka's published article long before the release of the CIA materials.[3]

Curiously, the documents were downgraded from "secret"

to "confidential" in March 1976. Two years later "Turner"—presumably Stansfield Turner, CIA chief from 1977 to 1981—exempted them from automatic declassification with a review date of 2020. Was the lumbering CIA bureaucracy really going to wait until 2020 to declassify confidential documents? Or did the flak surrounding the Church Committee reports published in 1975–76 and the revelation about LSD experiments (and the MKULTRA project) compel the CIA to guard its crown jewels more closely? It's not the secrets in the documents that make this release important but rather its illustration of the CIA's arbitrary and ridiculous information-sharing policies.

In fact, the NSA material on 895 secret-ink tests is much more revealing about secret-ink methods than that from the CIA. Keeping these files closed on national security grounds might be more justified. But the fact remains that if the CIA had released the material earlier, it would in no way have hurt national security or provided building blocks for current methods. First, the principle behind the inks was already available in published material; second, spy agencies had already developed much more sophisticated systems and methods.

Classifying this technical knowledge does demonstrate, however, that spy agencies found invisible ink a sensitive and important tool. It was certainly more widely used during the twentieth century than many people would imagine. In fact, as I hinted in the preface to the book, some people think the subject of secret ink is not a serious one—that it is all fun and games and magic tricks. Spy agencies certainly didn't see it this way. It was perceived as a valuable method of communication: no secret communication, no espionage. On this count they are right.

And as we step back and reflect on the broad sweep of history from ancient to modern times, the story of secret writing shows that in times of war, there was an increase in the development and use of secret writing by spies and prisoners of war. In times of peace, prisoners and lovers needed the method as a means of concealment or escape, and scientists and entertainers were enchanted by its magic.

And beyond the practical applications, secret-communication methods were often deeply embedded in the society and culture in which they were created. Every time there was a major advancement in invisible secret writing, there was a concomitant major, if not cataclysmic, event in society. Think back to the clash between the ancient Greeks and the Persians, to the English Civil War, and especially to the world wars and the Cold War. Jean Hellot's discovery of cobalt sympathetic ink took place in a time of comparative peace, but during a period of upheaval in science and technology. Developments in chemistry in the nineteenth century and its popular dissemination led to the development of the "missing link" of the chemistry set that helped introduce magical color changes in children's basement labs and their imaginations—and their parents'.

As this book was going to press, Edward J. Snowden leaked explosive information about the National Security Agency's global mail interception abilities and efforts. During the Cold War, the CIA was in the limelight, while the NSA hid in the shadows and was dubbed "No Such Agency." After the Snowden revelations, the NSA took the place of the CIA as the center of attention of abuse of power, technological prowess, and mail interception. It may be more appropriate to dub the NSA "Not Secret Anymore." In contrast to treatment of the CIA in the 1970s, however, there has been no Frank Church to grill NSA leaders, nor has a congressional committee put the brakes on illegal actions, including all-encompassing mail interception and reading.

During two world wars, the United Kingdom and the United States scrutinized their citizens' mail to an extent that would be hard to justify during peacetime. Then the Cold War laid the

foundations for our current national security state, as bureau-
cratic spy agencies were founded, grew, and thrived. After 9/11
and the declaration of a war on terror, the border between wartime and peacetime blurred. In particular, the National Security
Agency received rich funding in a time of recession. It went on
a hiring and building spree. Almost five million Americans now
have security clearance. The NSA built new buildings across the
United States to store its newly acquired metadata. In an unprecedented historical step, it even built the Utah Data Center, a
facility about five times the size of the U.S. Capitol.[4] Its building
boom alone indicates that it plans to stay in place for a long time
collecting data on U.S. citizens and foreign suspects.

U.S. leaders could learn quite a bit about the rise and fall of
secret communication efforts throughout history. In less than
eighty-four years the ethos has changed from "gentlemen do
not read each other's mail" to "we have the power to read everyone's e-mail with one keystroke." What does it say about the
United States' new national security state and democracy when
there is no Church Committee to scrutinize wrongdoing and
illegal mail interception? What is the cost of security, in dollars
and in freedom? There will always be secrets—personal secrets,
political secrets, military secrets—and there will always be those
determined to uncover those secrets. For the United States in
the early twenty-first century, the handwriting may be on the
wall—in invisible ink, waiting to be revealed.

APPENDIX

Fun Kitchen Chemistry Experiments

JASON LYE AND
KRISTIE MACRAKIS

Note: Some of the experiments listed below employ hazardous materials or materials that could become hazardous if used improperly. In the checking and editing of these procedures, every effort has been made to identify the handling of potentially dangerous materials, and safety precautions have been inserted where we deem them appropriate. However, should the reader conduct any of these experiments, he or she does so at his or her own risk. The authors and publisher do not warrant or guarantee the safety of individuals using these procedures and specifically disclaim any and all liability arising directly or indirectly from the use or application of any information contained in this appendix. Should you consider attempting any of these experiments, please make sure you know the proper safety precautions and understand the risks. [Disclaimer added by Yale University Counsel]

Beautiful chemical color changes are often what captures the imagination of budding young chemists. As we examined a 1952 Chemcraft chemistry set together, we lamented the decline of this educational toy. We wondered whether the possible downside to making toys safer was that society had lost a steady stream of enthusiastic new scientists.

In this spirit, we describe some safe historical sympathetic ink experiments that we undertook with household materials. It was a lot of fun. If you do try these experiments, make sure to work with a professional chemist or take a safety tutorial; most of the experiments are as harmless as using fresh lemons and lemon juice in your kitchen or cooking—but of course even a squirt in the eye with lemon juice can be painful.

Doin' Porridge

One of my favorite invisible-ink stories, that didn't make the book, is about a Thomas Robins, imprisoned in Britain's Wandsworth Prison, who developed invisible ink from oatmeal, or porridge, as it is called in Britain, in 1935. He squeezed a portion of porridge "through a piece of rag" and then used the liquid with an ordinary pen nib. Even though he didn't try to use it to escape from prison, he presented it to His Majesty's Principal Secretary of State for the Home Office in an effort to win an early release. Robins thought that if "this knowledge were to become public prison property, it would be of stupendous benefit to "Englands [sic] underworld."

In Britain, "Doin' porridge" is a common euphemism for serving a prison sentence. Who knew that the porridge served for breakfast at His Majesty's pleasure in the 1930s could in fact be used as secret (although quite stodgy) writing ink? Since prisoners got porridge without milk every day for breakfast, it was an easy substance to find. A simple tincture of iodine dropped on the message developed it into a beautiful blue color. When British specialists from MI5 and the Metropolitan Police experimented with this ink, they noted that it did not appear with a quartz ultraviolet lamp when used with water. It was a novel method for them. As long as no chunks made it onto the page, the prison guards would not detect it. If milk was added, though, detection was easier. As a result, MI5 and the police recommended adding milk to prisoners' porridge as a preventative. Robins was not given an early release.[1]

It's the starch in the porridge that picks up the iodine and makes a blue-black-colored complex. Starch and iodine are extremely sensitive mutual indicators. Starch molecules look similar to long coiled springs; the deep blue color of the starch-iodine complex is thought to be due to iodine molecules fitting into the spaces between those coils.

We used the following recipe to make a starch ink in our kitchen:

Heat 2 cups of water to boiling in a small saucepan. Mix 2 heaping teaspoons of a starch (like oatmeal) in ⅓ shot glass of cold water, and stir to make a slurry.

Once the water in the saucepan is boiling, carefully add the white starch slurry, stirring and continuing to boil for two minutes to make a good solution. The starch solution will thicken as it cools. A message can be written with a fine paintbrush or a Q-tip. For a finer line, snip one end from the Q-tip and use the cut end.

Once dried, the writing is developed with a dilute iodine solution. Always wear gloves and glasses to protect skin and eyes. Simply brush medicinal iodine tincture (or povidone) on the paper or apply it with an eye dropper. The paper will also pick up iodine initially—but without the starch, most of the iodine will leave the paper within a few hours, really contrasting the blue starch writing.

Barfly Fluorescence

Have you noticed that your gin and tonic glows greenish-blue under the blacklights at bars? That's because tonic water contains quinine. Quinine is colorless to the human eye, but it absorbs ultraviolet light and then reemits it as blue visible light. We call this phenomenon fluorescence.

Use tonic water as an invisible ink on brown paper bags, manila file folders, and other nonwhite papers. Writing on white copier paper usually doesn't work, because most copy papers are brightened with fluorescent brightening agents—so the glow of the quinine is drowned out by the glow of the brighteners. A blacklight bulb can be purchased in most hardware, drug, or grocery stores, enabling one to read the letters in a darkened room.

Hansen, the Nazi "tooth spy," used quinine secreted in a tooth as a secret-ink pencil, but his quinine was a gummy substance placed in a tiny bag.

Juice Inks: Develop with Heat

Many common fruit and vegetable juices can be used directly, by writing on paper with a Q-tip, a match, or a toothpick. Indeed, as we learned in Chapter 2, orange juice sympathetic ink was used by John Gerard to escape from the Tower of London in the sixteenth century. Even though World War I German spies got caught using lemon juice, if no one suspects you have written a message, no one is likely to heat the paper.

Common kitchen liquids that can be used as sympathetic inks include lemon, lime, and orange juices, vinegar, and milk. Onion juice also works. The secret writing can be made visible by heating the paper, for instance, by carefully holding the page in a toaster oven or over a grill or a candle. Always be careful not to overheat or ignite the page! As the paper warms, the ink chars the paper to a dark brown.

Chemical Inks: Develop with Water

Grandma's pickling alum could certainly tell a tale of secret lovers and spies. You'd be surprised . . . And no doubt Queen Mary and her conspirators all wished that she had used an invisible ink that Queen Elizabeth I's Lord Walsingham didn't know about . . .

We mixed ¼ teaspoon of household alum into ⅓ shot glass of city water to solution. This invisible ink can be used to write on regular paper. Once dried, the message can be revealed—only transiently—when the paper is placed in a bowl of water. As the water soaks into the paper, the paper appears to turn darker as it absorbs the water—except where the alum was used to write the message, which now appears whiter. As time passes, the alum dissolves and the message disappears. A truly self-destructing secret communication.

Old Black Inks: Iron Sulfate– and Tannin-Based Invisible Inks

Did you know that there is enough iron in the human body to make a three-inch nail? It is an essential nutrient for most species. It's also something that can be used as a sympathetic ink. Even though iron sulfate is of low toxicity, it still stains the skin and makes rust stains on most things over time. For centuries, iron sulfate was combined with tannic acid extracted from oak "gallnuts" or "nutgalls" (called "oak apples" in Europe) to make black ink. While the gallnut is one of the best sources of tannins, several common food items also contain tannins—such as black tea and red wine.

Where is iron sulfate sold today? Sometimes you can find the pure crystals online; it can be used to help fertilize lawns, turning grass a lush dark green. Alternatively, iron sulfate tablets can be bought at most drugstores as a dietary supplement. However, some purification steps are necessary for store-bought pills.

First, wearing gloves and glasses, we rubbed off the coating on the surface of the tablets with a damp paper towel. Next, we dissolved the tablet in water and added 3–4 cleaned tablets to a shot glass ⅔ full of water and swirled. We let it sit for 20 minutes to allow the insoluble materials in the tablet to settle onto the bottom of the glass. The top layer containing the iron sulfate can optionally be used as is, by dipping with a Q-tip, or alternatively, the clear, pale green top layer can be decanted into a separate shot glass for further use.

This pale green iron (II) sulfate ink can be used in as a sympathetic ink and developed in lots of different ways to produce different colors.

Black tea can be used to develop iron ink to make a black color. Just make a strong cup of hot tea with two teabags. After the tea has steeped for a few minutes, use this liquid to develop the iron ink. The iron in the ink combines with the tannic acid in tea to make a black material. Similarly, tannin solution—purchased from a winemaking shop—will also make an intense black color with iron ink on the paper. Because of the tannins in red wine, this will also develop iron invisible ink.

This ink is reversible, as we saw in the chapters about the ancient Greeks, della Porta, and British and American spies during the American Revolution. You can use iron sulfate as the ink and develop it with gallnuts—or the other way around.

Ink to Pink

Phenolphthalein was a major ingredient in laxatives until recently. You can still buy phenolphthalein solution from biodiesel supply companies. It's usually sold as a solution in alcohol, so be sure to keep it away from open flames and other sources of ignition. Phenolphthalein makes an intense magenta color in the presence of strong bases. Here you have another secret ink that the Nazis used to impregnate a handkerchief during World War II. You can use the solution to write a letter on paper.

Once dried, the secret writing is revealed by exposing the paper to ammonia vapor. The paper can also be developed in a bath of either washing soda (sodium carbonate) or dilute ammonia. If you have asthma or any other respiratory problem, be sure to use the washing soda, as the ammonia might aggravate your condition.

Assuming that you can tolerate the ammonia, you can use an "ammonia jar" to expose the secret letter to vapors by putting a little ammonia cleaner in the bottom of a quart-sized jar, then wedging the letter in the jar above the liquid and putting the lid on. The writing will appear in pink as the ammonia fumes permeate the paper. Alternatively, you can dilute 1 part ammonia in 4 parts water and pour that into an old plate in your sink to make a developer bath.

Chemical Inks: Cobalt Chloride, Develop with Heat

Years ago, chemistry sets contained all kinds of chemicals that nowadays can be procured only at chemical supply houses. One of the prettier chemicals included in the list was cobalt chloride, which forms beautiful ruby red crystals in its fully hydrated form and turns a bright blue color upon dehydration. It is that same color change that fascinated Jean Hellot and the Victorians, who used cobalt compounds to make changeable-landscape fire screens.

Occasionally, you can find an old 1960 chemistry set at a flea market or antique store that may still contain some fifty-year-old cobalt chloride. We conducted this experiment using chemistry set cobalt chloride once and a chemical supply house batch another time. They both worked beautifully. While cobalt chloride is pretty, it is also extremely toxic, a suspected cancer-causing agent, and really bad for the environment, so we were really careful, using gloves and goggles as laboratory safety protocols require. We do not recommend imitating us unless you are working in an industrial or university laboratory under direct supervision.

Making a cobalt ink can be as simple as dissolving a few of the beautiful ruby red crystals in tap water (while wearing gloves and safety glasses, of course). We dissolved a few red crystals (about ⅛ teaspoon) in ¼ shot glass of water to make a pink solution. In a hardwater area, we would have used distilled or deionized water instead of city water.

Aware of all the dangers, we then had a little fun. First, we drew a tree and landscape using a felt-tipped marking pen. Then we used a Q-tip, with the cotton puff removed to make a sort of stylus, to draw the leaves on the tree and a secret message.

Once dried, the pink color is too faint to see with the unaided eye.

(You can see it with a blacklight, though.) When warmed either in front of a fire or by a hair dryer or other dry heat source, however, the cobalt ink appeared in blue, creating the changeable landscape and revealing the secret message. When removed from the heat, the blue writing and images faded as the ink absorbed moisture from the air and rehydrated the cobalt.

The Wonderful Magical Talisman

A similar parlor game uses harmless heat-activated inks, so the experimenter does not have to worry about explosions or other threats to her health.

Make a little triangular box about 5 inches on each side and divided into three parts on the inside. The box must have a bottom (part A), a middle case (part B), and a top (part C) that covers the middle section. Place a copper plate on the bottom of the box, on which you have placed hieroglyphic characters contiguous to each other made out of different kinds of metals. Place a knob on the top part so that you can fit it on part B. Place another copper plate on the bottom of part B.

Decorate the outside of the talisman with unusual figures or characters to make it look more mysterious. Then write questions in common ink on pieces of paper, and on each write the answer with a sympathetic ink that will appear when heated. To make it more colorful when developed, you can write each word of the answer with a different sympathetic ink.

After you have heated the triangle and placed it under the cover, you can introduce the talisman and ask anyone from the party to choose a piece of paper with a question and place it in the talisman. Once the piece of paper is placed in the talisman and the cover placed on the top, the heat in the triangle will make the answer visible within a few minutes.

Some of the materials you can use in the talisman include standard heat-activated liquids like vinegar or citrus or onion juice—ingredients used in cookbooks and the kitchen.[2]

Hard-Boiled Egg Secret Messages

Phoo! Has anyone out there ever gotten this one to work? As you may recall, della Porta recommended using hard-boiled eggs to smuggle secret messages out of prison during the Inquisition. Evidently, messages were written on the shells with vinegar and alum as an ink. Supposedly, once the egg was peeled, the message could be read on the surface of the cooked egg. (We've tried writing on a raw eggshell, which we then boiled, with no success.)

With the eighth saucepan of eggs boiling away on the stove, the kitchen all steamed up, and eggshell crumbs crunching under our feet, we have tried all kinds of variations on this experiment, without success. Assuming it's not a complete myth, it seems as though some details of this secret-writing technique have been lost over the years. Also surprising is the number of references to the story that repeat the same incorrect procedural information almost verbatim— apparently without the authors' having confirmed that it works. If any of you has had any luck, please do reach out to us and let us know how to make it work.

Medicinal Messages

We learned that spies during the world wars were trained to use everyday medicines as invisible inks. While some of the medicines used in the 1910s and 1940s (like pyramidon) are no longer available, aspirin (acetyl salicylate) can still be found in most households.

Modern aspirin powders can be used to make invisible ink by placing a sachet of headache powder in a shot glass and adding hot water. This will make a saturated solution; as it cools, you will notice fluffy crystals coming out of the solution.

Write your secret message with the clear liquid and let it dry. Aspirin can be developed using fresh iron sulfate solution (see above). Dab it onto the dried page with a cotton pad or paper towel, and after a few minutes, the letters will begin to appear as a dark violet or sometimes brownish color. Color intensity continues to develop over time as the paper dries.

Safety

Even though we are talking about chemistry that can be done with household products, we followed strict safety instructions, and you should too. We remind you that we are not responsible for your safety, and we urge you, if you try any of these experiments at your own risk, to be sure to take general safety precautions; for example:

1. Personal protective equipment: Always wear gloves and safety glasses to protect your eyes. You can buy these at hardware stores.

2. Never use chemicals if you are pregnant, or think that you might be pregnant.

3. Minors should always be supervised by an adult.

4. Adults: It's best to not work alone. Not only is it more fun to have a partner, but it's much safer too!

5. If possible, keep a bowl of clean water close by so that you can wash your face if need be. Keep long hair in a ponytail while experimenting—this simplifies washing the face in an emergency. If you get something in your eye, irrigate it with water from the tap. Have a friend take you to the emergency room or call 911 if discomfort continues.

6. Never eat, drink, or smoke while handling chemicals. Don't touch any part of your face or any unprotected part of your body without first washing your hands.

7. Some experiments require paper to be heated. Take care not to ignite the paper accidentally, and have a fire extinguisher close by just in case.

8. Don't use kitchenware for food once you've used it for chemicals. You can buy cheap secondhand shot glasses, spoons, and plates from a thrift store.

About Jason

Jason Lye, Ph.D., C.Col., runs Lyco Works Incorporated, an innovation consulting company (lycoworks.com). Jason holds a Ph.D. in dye and pigment chemistry (NCSU), is a colour chemist (Leeds, England), and is a chartered colourist. He has had industrial experience at I.C.I., Accordis, and Kimberly-Clark and is a former director of external business development–technology assets for Newell Rubbermaid Inc.

NOTES

PREFACE

1. The results are discussed in Kristie Macrakis, Elizabeth K. Bell, Dale L. Perry, and Ryan D. Sweeder, "Invisible Ink Revealed: Concept, Context, and Chemical Principles of 'Cold War' Writing," *Journal of Chemical Education* 89 (2012), 529–32; Kristie Macrakis, *Seduced by Secrets: Inside the Stasi's Spy-Tech World* (New York: Cambridge University Press, 2008), 205–10. See also a recent book on technical steganography by Klaus Schmeh, *Versteckte Botschaften* (Munich: Heise, 2009).

1
THE ART OF LOVE AND WAR

1. Ovid, *The Art of Love, and Other Poems,* trans. J. H. Mozley, Loeb Classical Library (New York: Putnam, 1929), 162–63; Ovid *The Art of Love,* trans. James Michie (New York: Random House, 2002), 156–57.

2. Ibid., Mozley for translation.

3. Mozley, Introduction to *The Art of Love, and Other Poems,* x–xi.

4. Decimus Magus Ausonius, *Works,* trans. Hugh G. Evelyn White (New York: Putnam, 1921), 111.

5. Pliny, *Natural History,* trans. W. H. S. Jones (Cambridge: Harvard University Press, 1966), vol. 7, book 26, pp. 310–11.

6. Robert B. Strassler, ed., *The Landmark Herodotus: The Histories* (New York: Pantheon, 2007), p. 381; Polyaenus, *Strategems of War,* 2 vols., trans. P. Krentz and E. Wheeler (Chicago: Ares, 1994), 1: 49.

7. John Wilkins, *Mercury; or, the Secret and Swift Messenger* (London: I. Norton, 1641), 31.

8. Strassler, *The Landmark Herodotus,* 62–70.

9. Ibid., 68–71.

10. Ibid., 594–95.

11. See artists' representation of the battle of Salamis in Philip de Souza, ed., *The Ancient World at War: A Global History* (New York: Thames and Hudson, 2008), 103.

12. Adrienne Mayor, *The Poison King: The Life and Legend of Mithradates* (Princeton: Princeton University Press, 2010), 13.

13. Ibid., 196.

14. Ibid., 197.

15. Polyaenus, *Strategems of War*, 2 vols., trans P. Krentz and E. Wheeler (Chicago: Ares, 1994), 2: 1011–15; Aeneas Tacticus, "On the Defense of Fortified Positions," *Aeneas Tacticus, Asclepiodotus, Onasander,* trans. Illinois Greek Club (Cambridge: Harvard University Press, 1923), 1–199; Aineius the Tactician, *How to Survive under Siege,* trans. David Whitehead (New York: Oxford University Press, 1990).

16. Mayor, *The Poison King,* 206–7. Thanks to Adrienne Mayor for calling my attention to the Sulla episode.

17. Aeneas Tacticus, "On the Defense of Fortified Positions"; 157–59.

18. Ibid.

19. Ibid.

20. M. Thévenot, J. Boivin, and P. de La Hire, eds., *Veterum mathematicorum opera* (Paris: Ex Typographia Regia, 1693), 102, my translation. See also H. Diels and E. Schramm, *Exzerpte aus Philons Mechanik b. VII und VIII. Abhandlungen der Preussischen Akademie der Wissenschaften* (Berlin: Verlag der Akademie der Wissenschaften, 1920).

21. David Carvalho. *Forty Centuries of Ink; or, a Chronological Narrative Concerning Ink and Its Backgrounds* (New York: Bank's Law Publishing, 1904), 34.

22. William Shakespeare, *Twelfth Night,* act III, scene 2, numerous editions.

23. David Kahn, *The Codebreakers: The Comprehensive History of Secret Communication from Ancient Times to the Internet,* rev. and updated (New York: Scribner, 1996), 73.

24. Joseph Needham et al., *Science and Civilisation in China,* vol. 5, part 1, by Tsien Tsuen-Hsuin (Cambridge: Cambridge University Press, 1985), 247. See also ibid., vol. 5, part 4, p. 315.

25. Ibrahim A. al-Kadi, "Origins of Cryptology: The Arab Contributions," *Cryptologia* 16, no. 2 (1992), 97–126; for quotation, see Kahn, *The Codebreakers,* 93.

26. Al-Kadi, "Origins of Cryptology." For newer work, see Kathryn A. Schwartz, "Charting Arabic Cryptology's Evolution," *Cryptologia* 33 (2009), 97–304.

27. Siegfried Türkel, "Eine orientalische sympathetische Tinte im Mittelalter," *Archiv für Kriminologie* 79 (1926), 166.

28. C. E. Bosworth. "The Section on Codes and Their Decipherment in Qalqashandi's Subh Al-A'sha," *Journal of Semitic Studies* 8, no. 1 (1963), 17–33; for translation of invisible ink, see 22–23.

29. H. J. Webber. "History and Development of the Citrus Industry," in *The Citrus Industry,* 2nd ed., ed. W. Reuther, L. D. Batchelor, and H. J. Webber (Berkeley: University of California Press, 1967), 1–39; Pierre Laszlo, *Citrus: A History* (Chicago: University of Chicago Press, 2007).

30. See William Eamon, *Science and the Secrets of Nature: Books of Secrets in Medieval and Early Modern Culture* (Princeton: Princeton University Press, 1994).

31. John B. Friedman, "Safe Magic and Invisible Writing in the Secretorum Philosophorum," in *Conjuring Spirits: Texts and Traditions of Medieval Ritual Magic,* ed. Claire Fanger (University Park: Pennsylvania State University Press, 1998), 76–86.

2
INTRIGUE AND INQUISITION

1. Louise George Clubb, *Giambattista Della Porta: Dramatist* (Princeton: Princeton University Press, 1965), xi.

2. Giambattista della Porta, *Magia naturalis* (Naples, 1558, expanded 1589); John Baptista Porta, *Natural Magick* (London: John Wright, 1658, 2nd ed. with added book on invisible writing).

3. David Kahn, *The Codebreakers: The Comprehensive History of Secret Communication from Ancient Times to the Internet,* rev. and updated (New York: Scribner, 1996), 143.

4. Ioan. Baptista Porta, *De furtivis literarum notis* (Naples: Scotus, 1563); Giambattista della Porta, *On Secret Notations for Letters Commonly Called Ciphers,* trans. Keith Preston, 1916, unpublished, but available at Huntington Library, 41, 45.

5. William Eamon, *Science and the Secrets of Nature: Books of Secrets in Medieval and Early Modern Culture* (Princeton: Princeton University Press, 1994), 200.

6. Clubb, *Giambattista Della Porta,* 13.

7. Ibid.

8. Ibid., 54; Eamon, *Science and the Secrets of Nature,* 202.

9. Della Porta, *Natural Magic,* book 16, chapter 4.

10. Girolamo Cardano, *De rerum varietate, libri XVII* (Avignon: M.

Vincentius, 1558); for a German translation see Girolamo Cardano, *Offen-barung der Natur und natürlicher Dingen auch mancherley subtiler Würck-ungen*, trans. Heinricus Pantaleon (Basel: Sebastian Henricpetri, 1559).

11. Porta, *De furtivis literarum notis.*

12. "Message on an Egg Is Tempting to a Child," *New York Times*, May 29, 1965.

13. Kahn, *The Codebreakers*, 109.

14. "Venice: March 1531," March 4, Sanuto Diaries, Calendar of State Papers Relating to English Affairs in the Archives of Venice, vol. 4, 1527–1533, pp. 277–78, British History Online, www.british-history.ac.uk. See also Aloys Meister, *Die Anfänge der modernen Diplomatischen Geheimschrift* (Paderborn, Germany: Ferdinand Schöninger, 1902), 22; David S. Katz, *The Jews in the History of England, 1485–1850* (Oxford: Oxford University Press, 1994), 25.

15. Verini's book is available at the Newberry Library: Giovanni Battista Verini, *Secreti et modi bellissimi nuovamente investigati per Giovambatista Verini Fiorentino: e professore de modo scribendi* (Florence, c. 1530–35). Stanley Morison, "Some New Light on Verini," *Newberry Library Bulletin* 3, no. 2 (1953), 41–45.

16. Alan Haynes, *Walsingham: Elizabethan Spymaster and Statesman* (Stroud, Gloucestershire: Sutton, 2004), 3–6.

17. See Robert Hutchinson, *Elizabeth's Spymaster* (New York: St. Martin's, 2006); Stephen Budiansky, *Her Majesty's Spymaster: Elizabeth I, Sir Francis Walsingham, and the Birth of Modern Espionage* (New York: Viking, 2005); Fletcher Pratt, *Secret and Urgent: the Story of Codes and Ciphers* (New York: Bobbs-Merrill, 1939), 77.

18. Budiansky, *Her Majesty's Spymaster.*

19. "Walsingham, Sir Francis," *The International Cyclopaedia: A Library of Universal Knowledge* (New York: Dodd, Mead, 1885), 15: 217.

20. John Guy, *Queen of Scots: The True Life of Mary Stuart* (New York: Houghton Mifflin, 2004), 21, 368, 465.

21. Budiansky, *Her Majesty's Spymaster*, 106–8.

22. Mary, Queen of Scots, to Bishop of Glasgow, November 6, 1577, United Kingdom, State Papers, SP 53/10, f. 95, p. 245.

23. Mary, Queen of Scots, to Monsieur de Mauvissière, January 5, 1583/4, State Papers, SP 53/13.

24. Budiansky, *Her Majesty's Spymaster*, 117–18.

25. Ibid., 119–20.

26. Ibid., 119–20, 134.

27. Haynes, *Walsingham*, 148–50.

28. Quoted in Terry Crowder, *The Enemy Within: A History of Spies, Spymasters, and Espionage* (New York: Osprey, 2008), p. 64.

29. Mary, Queen of Scots, to Monsieur de L'Aubespine [Châteauneuf], January 30 1585/6, State Papers, SP 53/17, p. 205.

30. See Simon Singh, *The Code Book: The Science of Secrecy from Ancient Egypt to Quantum Cryptography* (New York: Random House, 1999), 1–45.

31. Ibid., 3.

32. Philip Ball, *Bright Earth: Art and the Invention of Color* (New York: Farrar, Straus and Giroux, 2001), 58–59; Charles Singer and Derek Spence, *The Earliest Chemical Industry: An Essay in the Historical Relations of Economics and Technology Illustrated from the Alum Trade* (London: Folio Society, 1948).

33. Hutchinson, *Elizabeth's Spymaster,* 279.

34. Thomas Rogers to Secretary Walsingham, August 11, 1585, State Papers, SP 15/29, f. 52; August 25, 1585, SP 15/29 f. 59; September 30, 1585, SP 15/29, f. 65.

35. Cotton MSS., Caligula C ix, f. 566, quoted in Richard Deacon, *A History of the British Secret Service* (London: Granada, 1969), 48.

36. Berden to Phelippes, undated, State Papers, 195, no. 75.

37. Alan Haynes, *The Elizabethan Secret Services* (Gloucestershire: Sutton, 2000), 49.

38. Arthur Gregorye to Walsingham, February 1586, State Papers, Harley 286 f. 78, p. 132.

39. John Gerard, *The Autobiography of a Hunted Priest,* trans. Philip Caraman (New York: Pellegrini, 1959).

40. John Morris, *Father John Gerard, of the Society of Jesus,* 3rd ed. (London: Burns and Oates, 1881), 166.

41. Ibid., 241.

42. Gerard, *Autobiography,* 116–17.

43. Ibid., 119.

3
CONFESSING SECRETS

1. G.B., *Rarities; or, The Incomparable Curiosities in Secret Writing, Both Aswel* [as well as] *by Waters As Cyphers, Explained and Made Familiar to the Meanest Capacity: By Which Ministers of State May Manage the Intrigues of Court and Grand Concerns of Princes, the Ladies Communicate Their Amours, and Every Ordinary Person (onely Capable of Legible Writ-*

ing) May Order His Private Affairs with All Imaginable Safety and Secrecy (London: Printed by J.G. for Nath. Brook, 1665), 18.

2. Barbara Shapiro, *John Wilkins, 1614–1672: An Intellectual Biography* (Berkeley: University of California Press, 1969), 1–11. For biographical information see also Hans Aarsleff, "John Wilkins," *Dictionary of Scientific Biography* (New York: Scribner, 1975), 14: 361–81.

3. Aarsleff, "John Wilkins"; Lisa Jardine, *On a Grander Scale: The Outstanding Life and Tumultuous Times of Sir Christopher Wren* (New York: HarperCollins, 2004), 128–29.

4. Andrew Clark, ed., *"Brief Lives," chiefly of Contemporaries, set down by John Aubrey, between the Years 1669 & 1696* (Oxford: Clarendon, 1898), 2: 300–301; Robert Chambers, "Dr. John Wilkins," *Chambers's Cyclopaedia of English Literature* (London: W. and R. Chambers, 1876), 339.

5. John Wilkins, "To the Reader," *Mercury; or, The Secret and Swift Messenger. Shewing, How a Man may with Privacy and Speed communicate his Thoughts to a Friend at any Distance,* 3rd ed. (1641; London: John Nicholson, 1707).

6. Ibid., 1641 ed., 30–31; 1707 ed., 16. For more on the tattooed, shaved head example see Kristie Macrakis, "Ancient Imprints: Fear and the Origins of Secret Writing," *Endeavour* 33, no. 1 (2009), 24–28. For the classic study on the quarrel between the ancients and moderns see R. F. Jones, *Ancients and Moderns: A Study of the Rise of the Scientific Movement in Seventeenth Century England* (1936; Berkeley: University of California Press, 1965).

7. John Wilkins, *Mercury* (1641), 179–80; (1707), 89–90.

8. Dorothy Stimson, "Dr. Wilkins and the Royal Society," *Journal of Modern History* 3, no. 4 (1931), 539–63; Margery Purver, *The Royal Society: Concept and Creation* (Cambridge: MIT Press, 1967).

9. James Gerald Crowther, *Founders of British Science* (London: Cresset, 1960), 145; Robert Theodore Gunther, *Early Science in Oxford* (Oxford: Clarendon, 1923), 78: 392.

10. Clark, *"Brief Lives,"* 2: 147; Thomas Birch, *The History of the Royal Society of London for Improving of Natural Knowledge, From its First Rise* (London: Printed for A. Millar in the Strand, 1756), 2: 24.

11. Birch, *History of the Royal Society,* 2: 345, 3: 179. For priority dispute see Adrian Johns, *The Nature of the Book: Print and Knowledge in the Making* (Chicago: University of Chicago Press, 1998), 522–23.

12. Birch, *History of the Royal Society,* 3: 138.

13. Marie Boas Hall, *Henry Oldenburg: Shaping the Royal Society* (Oxford: Oxford University Press, 2002).

14. Douglas McKie, "The Arrest and Imprisonment of Henry Olden-

burg," *Notes and Records of the Royal Society of London* 6, no. 1 (1948), 28–47.

15. Ibid.

16. Malcolm Oster, "Virtue, Providence, and Political Neutralism: Boyle and Interregnum Politics," in *Robert Boyle Reconsidered,* ed. Michael Hunter (Cambridge: Cambridge University Press, 1994), 19–36.

17. Clark, *Brief Lives,* 121.

18. See Charles Webster, "New Light on the Invisible College: The Social Relations of English Science in the Mid-Seventeenth Century," *Transactions of the Royal Historical Society* 24 (1974), 19–42; Clark, *Brief Lives,* 166. For biographical information: L. T. More, *The Life and Works of the Honourable Robert Boyle* (London: Oxford University Press, 1944).

19. Lawrence Principe, "Robert Boyle's Alchemical Secrecy: Codes, Ciphers, and Concealments," *Ambix* 39, no. 2 (1992), 63–74.

20. Robert Boyle, *Certain Physiological Essays and other Tracts Written at Distant Times* (London: H. Herringman, 1669), 35.

21. Ibid.

22. See Early English Books Online, keyword search "invisible ink" and "sympathetic ink," http://eebo.chadwyck.com. See also *Oxford English Dictionary,* as well as its online entry on invisible ink, http://dictionary.oed .com. Robert Boyle, *New Experiments and Observations Touching Cold; or, An Experimental History of Cold . . .* (London: John Crook, 1665), 844–45 ("A Postscript"). In addition to these references to invisible ink in his published writing, Boyle's papers refer to invisible ink. See Robert Boyle Papers, The Royal Society, 25: 367–85. "For the good of mankind" quotation: Eustace Budgell, *Memoirs of the Lives and Characters of the Illustrious Family of the Boyles* (London: Olive Payne, 1737), 144–45.

23. Robert Boyle, *The Philosophical Works of the Honourable Robert Boyle,* ed. Peter Shaw (London: Longman, 1725), 3: 468.

24. Boyle, *New Experiments and Observations,* 844–45.

25. Nicolas Lemery, *Cours de Chymie* (Paris: D'Houry, 1757); English translation: *Course of Chemistry* (London: Kettilby, 1686); Owen Hannaway. "Nicolas Lemery," *Dictionary of Scientific Biography* (New York: Scribner, 1973), 8: 172–75.

26. Lemery, *Cours de Chymie,* 330–33; *Course of Chemistry,* 260–63.

27. Pierre Borel, *Historiarum, et observationum medico-physicarum, centura prima: in qua, non solum, multa utilia, sed & rara, stupenda ac inaudita continentur* (Castries: Apud Arnaldum Colomerium, Regium Typographum, 1653); English translation in John Beckman, *A History of Inventions and Discoveries,* 2nd ed., trans. William Johnstone (London: Walker, 1814), 1:

175–76; Eduard Farber, "Pierre Borel," *Dictionary of Scientific Biography,* 2: 305–6; "Peter Borel," *Chalmers Biography* (London: J. Nichols and Son, 1812), 6: 106.

28. A. Rupert Hall and Marie Boas Hall, eds., *The Correspondence of Henry Oldenburg,* vol. 1, *1641–1662* (London: Taylor and Francis, 1965), 136–39.

29. Henry Baker, *Employment for the Microscope. In Two Parts: Likewise a Description of the Microscope Used in These Experiments* (London: Printed for R. Dodsley, 1753), 135. One example from 1890: "Invisible Inks," *English Mechanics and the World of Science* 51, p. 363.

30. John Davys, *An Essay on the Art of Decyphering. In which is inserted a Discourse of Dr. Wallis* (London: L. Gilliver and J. Clarke, 1773), 13; Christopher J. Scriba, ed., "The Autobiography of John Wallis, F.R.S.," *Notes and Records of the Royal Society of London* 25, no. 1 (1970), 37–38; David Eugene Smith, "John Wallis as Cryptographer," *Bulletin of the American Mathematical Society,* 1917, pp. 83–84.

31. Smith, "John Wallis as Cryptographer." Wallis's letters reprinted p. 83. Davys, *Essay on the Art of Decyphering,* p. 30.

32. Wallis manuscript BL Add. MSS 32499, fo. 377, quoted in Alan Marshall, *Intelligence and Espionage in the Reign of Charles II, 1660–1685* (Cambridge: Cambridge University Press, 1994), 93; Smith, "John Wallis as Cryptographer."

33. Smith, "John Wallis as Cryptographer," letter of August 15, 1691, p. 91.

34. David Kahn, *The Codebreakers: The Comprehensive History of Secret Communication from Ancient Times to the Internet,* rev. and updated (New York: Scribner, 1996), 167. See also Peter Pesic, "Secrets, Symbols, and Systems: Parallels between Cryptanalysis and Algebra, 1580–1700," *Isis* 88 (1977), 674–92.

35. Davys, *Essay on the Art of Decyphering,* 10.

36. J. A. Reeds, "Solved: The Ciphers in Book III of Trithemius's *Steganographia,*" *Cryptologia* 22 (October 1998), 291–319.

37. Ibid., 293.

38. Johannes Trithemius, *Polygraphiae,* book 6, *Argentoratum* (Zetzner, 1613), 50; John Falconer, *Cryptomenysis Patefacta; Or, the Art of Secret Information Disclosed Without a Key* (London: Daniel Browne, 1685), discussion of Trithemius, passim.

39. Paula Findlen, ed., *Athanasius Kircher: The Last Man Who Knew Everything* (New York: Routledge, 2004); Nick Wilding, "'If You Have a Secret, Either Keep It, or Reveal It': Cryptography and Universal Language," in *The Great Art of Knowing: The Baroque Encyclopedia of Athanasius*

Kircher, ed. Daniel Stolzenberg (Stanford: Stanford University Libraries, 2001), 99–103.

40. Wilding, "'If You Have a Secret,'" 102; Findlen, *Athanasius Kircher,* 292–93.

41. Haun Saussy, "Magnetic Language: Athanasius Kircher and Magnetic Language," in Findlen, *Athanasius Kircher,* 263–82.

42. Johannes B. Friderici, *Cryptographia Oder Geheime Schrifft-Münd Und Würckliche Correspondentz, Welche Lehrmässig Voratellet Eine Hoch-Schätzbare Kunst Verborgene Schrifften Zu Machen Und Auffzulösen, in Sich Begreiffend Viel Frembde Und Verwunderungs-Würdige Arten, Wie Man Durch Versetzung Der Buchstaben, Item Durch Allerhand Characteres, Ziffern, Noten, Punkte Seine Meinung Gewissen Personen, Gantz Verborgener Weise, Kan Zu Verstehen Geben* (Hamburg, 1685).

43. Adam Fitz-Adam, *The World, for the Year One Thousand Seven Hundred and Fifty Three* (London: R. and J. Dodsley, 1753), no. 24, p. 145. Most dictionaries up until 1910 used this definition. See, for example, *Webster's Practical Dictionary* (Chicago: Reilly and Britton, 1910), 413: "the art of writing in ciphers, or characters not intelligible except to the persons who correspond with each other." For the early modern period see Early Books Online, Eighteenth Century Online, and the *Oxford English Dictionary.* For the period since 1800 see www.ngramviewer.com: keyword "steganography" yields much activity between 1800 and 1825 and after the year 1900.

44. State Papers, Britain, Smith to Warwick, November 10, 1562.

45. National Archives, Britain, SP 105/58 Stepney to Paget, Vienna, March 22, 1693.

46. N. Bailey, *An Universal Etymological English Dictionary: Comprehending the Derivations of the Generality of Words in the English Tongue . . . and Also a Brief and Clear Explication of All Difficult Words . . . Together with a Large Collection and Explication of Words and Phrases Us'd in Our Ancient Statutes, Charters, Writs . . . and the Etymology and Interpretation of the Proper Names of Men, Women, and Remarkable Places in Great Britain . . . to Which Is Added, a Collection of Our Most Common Proverbs* (London: Printed for J. Darby . . . [et al.], 1728), entry for "Sympathetic Inks."

4
INVISIBLE LANDSCAPES

1. This description is based on a visit to Schneeberg in 2010 by the author; Schneeberg Chronik at the Museum.

2. Paul Strathern, *Mendeleyev's Dream: The Quest for the Elements* (New York: St. Martin's, 2000), 94; Ekkehard Schwab, "Cobalt," *Chemical and Engineering News*, 2003, http://pubs.acs.org/cen/80th/cobalt.html.

3. Peter Hammer, "Das Sächsische Blaufarbenwesen und der Handel mit Kobaltfarben—nach Unterlagen der Bücherei der Bergakademie Freiberg," in *VII International Symposium "Cultural Heritage in Geosciences, Mining, and Metallurgy: Libraries—Archives—Museums": "Museums and Their Collections," Leiden (The Netherlands), 19–23 May 2003*, ed. C. F. Winkler Prins and S. K. Donovan, *Scripta Geologica Special Issue* 4 (August 2004), 108–17.

4. H. C. Erik Midelfort, *Madness in Sixteenth-Century Germany* (Stanford: Stanford University Press, 2000), 52; Arthur Edward Waite, *Hermetic and Alchemical Writings of Paracelsus* (Whitefish, Mont.: Kessinger, 2005), 254.

5. See DJW, *Schlüssel zu dem Cabinet der geheimen Schatz-Kammer der Natur* (Leipzig: Verlegts J. Heinichs Witwe, 1706), 285–88. For one-paragraph descriptions of Dorothea Juliana Wallich or Walchin and her books, see "Dorothea Juliana Wallichia," in Schacher's *Diss. Hist. Crit. De feminis ex arte medica claris* (Leipzig: Lipsia Langenheim, 1738), 51; Johann Heinrich Zedler, *Zedlers Universal-Lexicon* (1747), bd. 52, p. 1107; Johann Christian Friedrich Harless, "Walch, or Walchin (D. J.)," *Die Verdienste der Frauen um Naturwissenschaft: Gesundheits and Heilkunke* (1830), 176; John Ferguson, *Bibliotheca Chemica: A Catalogue of the Alchemical, Chemical, and Pharmaceutical Books in the Collection of the Late James Young of Kelly and Durris* (Glasgow: J. Maclehose and Sons, 1906), 525–26; Hermann Kopp, *Die Alchemie in älterer und neuerer Zeit: Ein Beitrag zuu Culturgeschichte* (Heidelberg: Carl Winter's Universitätsbuchhandlung, 1886), 364.

6. Kopp, *Alchemie,* 364.

7. Johann Beckmann, *Physikalisch-ökonomische Bibliothek* (Göttingen: Vandenhoek und Ruprecht Verlag, 1799), 20: 559–60; G. E. Stahl, *G. E. Stahls zufällige Gedanken und nützliche Bedencken über den Streit von dem so genannten Sulphure* (Halle: In Verlegung des Waysenhauses, 1718), 249–52.

8. No correspondence or portraits of Hellot survive. The description is taken from Arthur Birembaut and Guy Thuillier, "Une source Inédite: Les cahiers du Chimiste Jean Hellot (1685–1766)," *Annales. Économies, Sociétes, Civilisations* 21, no. 2 (1966), 357–64, here p. 359.

9. Jean-Paul Grandjean de Fouchy, "Éloges de M. Hellot," *Histoire de L'Academie Royale des Sciences*, 1766, 312–36; Todériciu Doru, "Chimie Appliquée et Technologue Chimique en France au Milieu du XVIIIe Siècle: Oevre et Vie de Jean Hellot, Thèse de Troisième Cycle, École Practique des Hautes Études, Université de Paris-Sorbonne, 1975; Marie Boas Hall,

"Jean Hellot," in *Dictionary of Scientific Biography,* ed. Charles Gillispie et al. (New York: Scribner, 1972), 6: 236–37; Jaime Wisniak, "Jean Hellot: A Pioneer of Chemical Technology," *Revista CENIC Ciencias Químicas* 40, no. 2 (2009), 111–21.

The economic crisis was caused the disastrous application of John Law's theories. Law is said to be the father of finance because he introduced paper bills. Nevertheless, the note-issuing bank he created failed and caused a crisis in France and Europe.

10. Hall, "Jean Hellot." For a cogent description of *adjoint* and *pension-naires* titles at the Academy, see Roger Hahn, *The Anatomy of a Scientific Institution: The Paris Academy of Sciences, 1666–1803* (Berkeley: University of California Press, 1971), 78–79.

11. Wisniak, "Jean Hellot."

12. Grandjean de Fouchy, "Éloge de M. Hellot"; Marie Boas Hall, "Jean Hellot."

13. Jean Hellot, "Sur une nouvelle Encre Sympatique," part 1, *Mémoires de L'Académie Royale des Sciences,* 1837, 101–20.

14. For some samples of references to "Hellot's sympathetic ink" see James Cutbush, *The Philosophy of Experimental Chemistry* (Philadelphia: Isaac Peirce, 1813), 323; *The Penny Cyclopaedia of the Society for the Diffusion of Useful Knowledge* (London: Charles Knight, 1838), 12: 478. Even Germans started using the eponymous "Hellot's sympathetische Tinte": see Ernst Ludwig Schubarth, *Lehrbuch der theoretischen Chemie: zunächst für Aerzte und Pharmaceuten* (Berlin: Rücker, 1822), 304.

15. Hellot, "Sur une nouvelle Encre Sympatique," 101–2.

16. Johann Albrecht Gesner and Johnann C. Erhard, *Selecta physico-oeconomica, oder Sammlung von allerhand zur Naturfoschung und Haushaltungskunst gehörigen Begebenheiten* (Stuttgart, 1752).

17. Jean Hellot, "Seconde Partie du Mémoire aur L'Encre Sympathique, ou Teinture Extraite des Mines de Bismuth, d'Azure and d'Arsenic," *Mémoires de L'Académie Royale des Sciences* (1837), 244–45.

18. John Talbot Dillon, *Travels through Spain, with a View to Illustrate the Natural History and Physical Geography of that Kingdom* (London, 1780), 219.

19. Mrs. Lincoln Phelps, *Chemistry, Collegiate Institutions, Schools, Families, and Private Students* (New York: Sheldon, Blakeman, 1857), 209.

20. Hans H. Neuberger and F. Briscoe Stephens, *Weather and Man* (New York: Prentice Hall, 1948), 129; Raymond B. Wailes. "Feats of Magic for the Home Chemist," *Popular Science Monthly,* September 1934, 61.

21. Pierre-Joseph Macquer, *Dictionnaire de chymie* (Paris: Lacombe,

1766); Pierre-Joseph Macquer, *A Dictionary of Chemistry* (London: T. Cadell, and P. Emsly, 1771); Edward Hussey Delaval, *An Experimental Inquiry into the Causes of the Changes of Colours in Opake and Coloured Bodies* (London: J. Nourse, 1777), 85. "Chemistry for chemistry's sake": see Jonathan Simon, *Chemistry, Pharmacy, and Revolution in France, 1777–1809* (Aldershot: Ashgate, 2005), from table of contents subchapter head.

22. Erasmus Darwin, *The Botanic Garden, a Poem. In Two Parts. Part I. The Economy of Vegetation. Part II. The Loves of the Plants,* 4th ed. (London: J. Johnson, 1799), 56–57.

23. Ibid., Preface.

24. Ibid., 30, on "zaffre, as sold by the druggists."

25. Richard Holmes, *The Age of Wonder: How the Romantic Generation Discovered the Beauty and Terror of Science* (New York: Vintage, 2010).

5
REVOLUTIONARY INK

1. United States House of Representatives, *Report from the Committee to whom was referred on the fourth instant, the petition of James Jay, of the state of New-York* (Washington, D.C.: A. and G. Way, 1807). See also James Jay to an unnamed general, January 9, 1808, Special Collections, University Archives, Stony Brook, N.Y.

2. M. H. T., "James Jay," in *Dictionary of American Biography* (New York: Scribner, 1928–1956), 10: 4–5; Thomas Jones, *History of New York during the Revolutionary War* (New York: New-York Historical Society, 1879), 2: 223; James Jay to Thomas Jefferson, April 14, 1806, Thomas Jefferson Papers, series 1, General Correspondence, 1651–1827, Library of Congress.

3. Jay to Jefferson, April 14, 1806; Victor Hugo Paltsits. "The Use of Invisible Ink for Secret Writing during the American Revolution," *Bulletin of the New York Public Library* 39 (1935), 361–64; quotation on 362.

4. Joel Richard Paul, *Unlikely Allies: How a Merchant, a Playwright, and a Spy Saved the American Revolution* (New York: Riverhead, 2009), 8–9; William Jay, *The Life of John Jay: With Selections from his Correspondence and Miscellaneous Papers* (New York: J. and J. Harper, 1833), 1: 64.

5. Paul, *Unlikely Allies,* 7–17.

6. Silas Deane to Robert Morris, September 17, 1776, Silas Deane and Charles Isham, *The Deane Papers . . . 1774–[1790]* (New York: Printed for the [New-York Historical] Society, 1887), 247.

7. John Jay to Robert Morris, September 23, 1776, first reprinted in William Jay, *The Life of John Jay,* 65; John Jay to Robert Morris, September

15, 1776, in the Papers of John Jay, New York Public Library; John Jay to Robert Morris October 6, 1776, first reprinted in William Jay, *The Life of John Jay,* 66.

8. Robert Morris to John Jay, February 4, 1777, first reprinted in William Jay, *The Life of John Jay,* 66.

9. Morton Pennypacker, *The Two Spies: Nathan Hale and Robert Townsend* (Boston: Houghton Mifflin, 1930); Morton Pennypacker, *George Washington's Spies on Long Island and in New York* (Long Island Historical Society, 1939).

10. Alexander Rose, *Washington's Spies: The Story of America's First Spy Ring* (New York: Bantam, 2006), 72.

11. Cory Ford, *A Peculiar Service* (Boston: Little, Brown, 1965), 162.

12. George Washington to Benjamin Tallmadge, December 17, 1778, George Washington Papers, Library of Congress.

13. John Jay to George Washington, November 19, 1778, Library of Congress; James Jay to Jefferson, April 14, 1806.

14. George Washington to Elias Boudinot, May 3, 1779, Washington Papers, Library of Congress.

15. Ford, *A Peculiar Service,* 163–64.

16. Rose, *Washington's Spies,* 131–32.

17. Thomas B. Allen, *George Washington, Spymaster* (Washington, D.C.: National Geographic, 2004), 55–57.

18. George Washington to Benjamin Tallmadge, July 25, 1779, Westchester Archives, original and transcript available at http://westchesterarchives .com/ht/muni/tarrytn/tallmadge.htm.

19. Ibid.

20. George Washington to Benjamin Tallmadge, September 24, 1779, Stony Brook University Libraries, Special Collection and Archives, Manuscript Collection 402.

21. George Washington to Benjamin Tallmadge, February 5, 1780, Benjamin Tallmadge Papers, Princeton University, Department of Rare Books and Special Collections at Firestone Library.

22. George Washington to James Jay, April 9, 1780, George Washington Papers, Library of Congress.

23. James Jay to George Washington, April 13, 1780; Washington to Jay, May 12 1780, George Washington Papers, Library of Congress.

24. L. Bendikson, "The Restoration of Obliterated Passages and of Secret Writing in Diplomatic Missives," *Franco-American Review,* 1937, 240–56; Kenneth W. Rendell Rare Books Catalogue for 1972.

25. Bendikson, "Restoration of Obliterated Passages."

26. Frank Thone, "The 'Black Chamber' of 1776," *Science News Letter,* July 3, 1937, 6–7, 14.

27. Sanborn Brown and Elbridge W. Stein, "Benjamin Thompson and the First Secret-Ink Letter of the American Revolution," *Journal of Criminal Law and Criminology* 40 (January–February 1950), 627–36.

28. Ibid.

29. John André letters in Spy Letters of the American Revolution, Clements Library, University of Michigan, www.clements.umich.edu/exhibits/online/spies/gallery.html. See also Allen, *George Washington, Spymaster.*

30. Letter quoted in Howard H. Peckham, "British Secret Writing in the Revolution," *Michigan Alumnus Quarterly Review* 44 (December 4, 1937), 126–31, quotation on 127. This letter, dated May 31, 1779, is also available in the Clinton Papers at the University of Michigan Rare Books Library, 59: 27. See also John A. Nagy, *Invisible Ink: Spycraft of the American Revolution* (Yardley, Pa.: Westholme, 2010), 32–33. Notwithstanding the title of Nagy's book, he has only one chapter on invisible ink.

31. Thomas J. Schaeper, *Edward Bancroft: Scientist, Author, Spy* (New Haven: Yale University Press, 2011), 2–3, 47–48.

32. United States Congress, House Committee to whom was referred the petition of James Jay, *Report from the Committee to whom was referred on the fourth instant, the petition of James Jay, of the state of New-York* (Washington: A. and G. Way, 1807); James Jay to Jefferson, April 14, 1806, note 3.

33. United States Congress, House Committee to whom was referred the petition of James Jay, *Report from the Committee to who was referred on the fourth instant, the petition of James Jay, of the state of New-York,* "Saturday, November 21. Sir James Jay." *Abridgement of the Debates of Congress, from 1789 to 1856.* November 1807, p. 620.

34. *Report of the Committee on the Petition of James Jay,* July 6, 1813 (Washington, [D.C.]: Roger C. Weightman, 1813; "Memorial of Sir James Jay," *Annals of Congress,* November 1807, 951–53; *House Journal,* March 2, 1808; *House Journal,* April 20, 1808.

6
MAGIC

1. Chemcraft Magic, *Chemcraft Magic: Give a Magic Show* (Hagerstown, Md.: Porter Chemical Company, 1952), 2, 19–20.

2. For the use of the term "secular stage magic" see Simon During, *Modern Enchantments: The Cultural Power of Secular Magic* (Cambridge: Harvard University Press, 2002).

3. Richard D. Altick, *The Shows of London* (Cambridge: Harvard University Press, 1978), 17–19; Peter Bowler and Iwan Rhys Morus, *Making Modern Science* (Chicago: University of Chicago Press, 2005), 369; Roy Porter, *Enlightenment: Britain and the Creation of the Modern World* (London: Allen Lane, 2000); Tom Standage, *A History of the World in Six Glasses* (New York: Random House, 2005), 168.

4. Lorraine Daston and Katherine Park, *Wonders and the Order of Nature, 1150–1750* (New York: Zone, 1998).

5. Bowler and Morus, *Making Modern Science,* 376.

6. Louis-Sébastien Mercier and Jeremy D. Popkin, *Panorama of Paris: Selections from Le Tableau de Paris* (University Park: Pennsylvania State University Press, 1999), 30; Michael R. Lynn, *Popular Science and Public Opinion in Eighteenth-Century France* (Manchester: Manchester University Press, 2006), 1.

7. Robert Darnton, *Mesmerism and the End of the Enlightenment in France* (Cambridge: Harvard University Press, 1968), 23.

8. Milbourne Christopher and Maurine Christopher, *The Illustrated History of Magic* (New York: Carroll and Graf, 2006), 82–96.

9. Henri Decremps, *Philosophical Amusements; or, Easy Instructive Recreations for Young People* (London: J. Johnson, 1790), 63–71.

10. See, for example, David Brewster, *Letters on Natural Magic* (London: Murray, 1832); Decremps, *Philosophical Amusements;* William Hooper, *Rational Recreations in which the principles of numbers and natural philosophy are clearly and copiously elucidated . . .*, 4th ed., corrected (London, 1794).

11. Edmé-Gilles Guyot, *Nouvelles Récréations physique et mathématique . . .* (Paris: Gueffer, 1769); Jacques Ozanam, *Récréations mathématiques et physiques, qui contiennent plusieurs problémes utiles et agréables, d'arithmétique, de géométrie, de fusique, d'optique, de gnominique, de cosmographie, de mécanique, de pyrotechnie, et de physique; avec Un Traité nouveau des horloges elementaires* (Paris: Chez Jean Jombert, 1696), translated as Jacques Ozanam, *Recreations Mathematical and Physical: Laying Down and Solving Many Profitable and Delightful Problems of Arithmetick, Geometry, Opticks, Gnomonicks, Cosmography, Mechanicks, Physicks, and Pyrotechny* (London: Bonwick, Freeman, Goodwin, Waltho, Wotton, Manship, Nicholson, Parker, Tooke, and Smith, 1708); During, *Modern Enchantments,* 83, 87.

12. William Hooper, M.D., *Rational Recreations in which the Principles of Numbers and Natural Philosophy are Clearly and Copiously Elucidated by a Series of Easy, Entertaining, Interesting Experiments . . .* (London: L. David, 1774), iv. Hooper mentions Guyot in passing three times.

13. Johann Christian Wiegleb, *Die natürliche Magie: aus allerhand belustigenden und nützlichen Kunststücken bestehend; mit Kupfern* (Berlin: Nicolai, 1779); Johann Samuel Halle, *Magie, oder, Die Zauberkräfte der Natur: so auf den Nutzen und die Belustigung angewandt worden* (Berlin: Bey Joachim Pauli, 1783); Christlieb Benedict Funk, *Natürliche Magie, oder, Erklärung verschiedner Wahrsager-und natürlicher Zauberkünste* (Berlin: Bey Friedrich Nicolai, 1783).

14. Reprint of Wiegleb's and Rosenthal's volumes: Johann Christian Wiegleb, Gottfried Erich Rosenthal, and Johann Heinrich Moritz von Poppe, *Wieglebs und Rosenthals gesammelte Schriften über natürliche Magie: 73 Exempel d. Schwarzen Kunst* (Michelsneukirchen: Carussell-Verlag, 1982). Wiegleb's autobiographical notes are in [Friedrich Christian] Stoeller, "Nekrolog Johann Christian Wiegleb," *Allgemeines Journal der Chemie* 4 (1800), 684–720, quotations on 689. On the apothecary-chemists in eighteenth-century Germany see Ursula Klein, "Apothecary-Chemists in Eighteenth-Century Germany," in *New Narratives in Eighteenth-Century Chemistry*, ed. Lawrence M. Principe (Dordrecht: Springer, 2007), 97–138.

15. Karl Hufbauer, *The Formation of the German Chemical Community, 1720–1795* (Berkeley: University of California Press, 1982), p. 26. The journal was F. Nicolai's *Allgemeine deutsche Bibliothek*.

16. Johann Christian Wiegleb, *A General System of Chemistry, Theoretical and Practical. Taken Chiefly from the German of M. Wiegleb*, trans. C. R. Hopson, M.D. (London, 1789), quotation from introduction, 1.

17. Wiegleb, *Die natürliche Magie*, 1–6.

18. Göttling Instruction Manual at the Smithsonian Institution, Washington, D.C.; W. A. Smeaton, "The Portable Chemical Laboratories of Guyton de Morveau, Cronstedt, and Göttling," *Ambix* 13 (1966), 84–91. See also Brian Gee, "Amusement Chests and Portable Laboratories: Practical Alternatives to the Regular Laboratory," in *The Development of the Laboratory*, ed. Frank A. J. L. James (New York: American Institute of Physics, 1989), 37–59.

19. Rosie Cook, "Chemistry at Play," *Chemical Heritage Foundation Magazine*, 2010, available online at http://www.chemheritage.org/discover/media/magazine/articles/28-1-chemistry-at-play.aspx.

20. Ibid. for the chemical cabinets.

21. Johann Nikolaus Martius, *Unterricht in der natürlichen Magie, oder zu allerhand belustigenden und nützlichen Kunststücken*, totally rev. Johann Christian Wiegleb (Berlin: Friedrich Nicolai, 1779), 1–2. See also Achim M. Klosa, *Johann Christian Wiegleb (1732–1800). Eine Ergobiographie der Aufklärung* (Stuttgart: Wissenschaftliche Verlagsgesellschaft, 2009), 176. On

the publisher's wish to impregnate pages with sympathetic ink: Wiegleb to Nicolai, November 30, 1785, Deutsches Museum, 3755.

22. Jean-Jacques Rousseau, *Confessions* (Hertfordshire: Wordsworth Editions Limited, 1996), 211; Ozanam, *Récréations Mathématiques;* 436; Ozanam, *Recreations Mathematical,* 455.

23. E. Stanyon, *Fire and Chemical Magic for Drawing Room and Stage Performances. Containing full Explanations of all the Latest and most Startling Chemical Color Change Tricks* (London: E. Stanyon, 1909), 2. First cited in Salim al-Gailini, "Magic, Science, and Masculinity: Marketing Toy Chemistry Sets," *Studies in History and Philosophy of Science* 40 (2009), 375.

24. Al-Gailini, "Magic, Science, and Masculinity," 375.

25. John D. Lippy Jr. and Edward L. Palder, *Modern Chemical Magic* (Harrisburg, Pa.: Stackpole, 1959), ix.

26. Ibid., vii.

27. Chemcraft Magic, *Chemcraft Magic,* 19–20.

28. Cook, "Chemistry at Play."

29. Ronald Clark, *The Man Who Broke Purple* (Boston: Little, Brown), 1977, 11–12.

30. Thomas Ollive Mabbott, ed., *Collected Works of Edgar Allan Poe* (Cambridge: Harvard University Press, Belknap, 1969–78), 3: 799; Simon Singh, *The Code Book: The Science of Secrecy from Ancient Egypt to Quantum Cryptography* (New York: Random House, 1999), 82–83.

31. Edgar Allan Poe, "The Gold-Bug," *18 Best Stories by Edgar Allan Poe,* ed. Vincent Price and Chandler Brossard (New York: Dell, 1965), 144–81, quotation on 169.

32. Ibid., 169–70.

33. Ibid., 154.

7
THE SECRET-INK WAR

1. Sir Basil Thomson, *The Story of Scotland Yard* (1936; Whitefish, Mont.: Kessinger, 2005), 230. According to MI5 records and Thomas Boghardt, *Spies of the Kaiser* (London: Palgrave Macmillan, 2004), Karl Friedrich Müller was a Baltic German. I refer to him as Carl Muller because that was the name the British used in their files.

2. TNA, Kew, WO 141/2/2, Report by W. Culver Jones Surgeon-Colonel, Honourable Artillery Company; Sidney Theodore Felstead, *German Spies at Bay* (New York: Brentano's, 1920), 51.

3. Tammy M. Proctor, *Female Intelligence: Women and Espionage in the*

First World War (New York: New York University Press, 2003). Figure of three thousand female censors is from C. Ainsworth Mitchell, "Mabel Beatrice Elliott Obituary," *Analyst* 69, no. 815 (1944), 33–34; Felstead, *German Spies at Bay,* 71.

4. TNA, Kew, *Report on Postal Censorship during the Great War, 1914–1919,* only in archives; Felstead, *German Spies at Bay,* 71; David Kahn, *The Codebreakers* (New York: Scribner, 1996), 513.

5. Nicholas Hiley, "Counter-Espionage and Security in Great Britain during the First World War," *English Historical Review* 101 (1986), 635–70, statistics on 640.

6. TNA, Kew, WO 141/2/2, Examination of Carl Muller, 15–18, Evidence by Arthur Francois Brys, 14. See also Leonard Sellers, *Shot in the Tower* (London: Leo Cooper, 1997), chapter on Muller, 43–63.

7. TNA, Kew, WO 141/2/2, Examination of Carl Muller, 20–28.

8. TNA, Kew, WO 141/2/2, Copies of translation of letters marked A.

9. Wim Klinkert, "A Spy's Paradise? German Espionage in the Netherlands, 1914–1918," *Journal of Intelligence History* 12 (2013), 12–35.

10. TNA, Kew, WO 141/2/2, Christine Emily Hahn, March 19, 1915, 42–43.

11. TNA, Kew, WO 141/2/2, Evidence by Carl Muller, 111.

12. TNA, Kew, WO 141/2/2.

13. Obituary, Charles Ainsworth Mitchell (1867–1948), *Analyst* 73, no. 863 (1948), 55–57.

14. TNA, Kew, WO 141/2/2, Statement of Evidence taken March 1915 from Charles Ainsworth Mitchell at the Tower of London, 36–37, and evidence, second day, 2–4.

15. See Sellers, *Shot in the Tower,* 59.

16. Ibid., 60–61. See also TNA, Kew, WO 141/2/2.

17. Basil Thomson, *Queer People* (London: Hodder and Stoughton, 1922), 133.

18. Sellers, *Shot in the Tower,* 63.

19. Felstead, *German Spies at Bay,* 25–32; Thomas Boghardt, *Spies of the Kaiser,* 101.

20. Boghardt, *Spies of the Kaiser,* 101; Felstead, *German Spies at Bay,* 25.

21. Boghardt, *Spies of the Kaiser,* 101.

22. Felstead, *German Spies at Bay,* 33–36.

23. TNA, KV 4/112, "Kupferle, Anthony," "Game Book," vol. 1, 1919–1915. See also Christopher Andrew, *The Defence of the Realm: The Authorized History of MI5* (London: Allen Lane, 2009), 67.

24. TNA, KV 4/112, "Kupferle, Anthony," "Game Book," vol. 1, 1919–1915. See also Andrew, *Defence of the Realm*, 67.

25. Mitchell, "Mabel Beatrice Elliott Obituary." See also TNA, KV 1/74, "The Testing Department: Short Report on work done during the War," 124–25.

26. Slate message reproduced in *The Times* (London), May 21, 1915, and Felstead, *German Spies at Bay*, 38–39.

27. See the article in *The Times,* May 21, 1915.

28. Felstead, *German Spies at Bay,* 152.

29. TNA, Kew, DPP1/42, p. 10 of secret-writing transcripts.

30. TNA, Kew, DPP 1/42, p. 7 of cross-examination of Miss Graham.

31. TNA, Kew DPP1/42, pp. 13, 24 of cross-examination of John Price Millington.

32. TNA, Kew, DPP 1/42.

33. TNA, Kew, *Report on Postal Censorship during the Great War, 1914–1919*, 405–7.

34. Felstead, *German Spies at Bay,* 76.

35. TNA, Kew, DEFE 1/115 for quotation; "Stanley Winter Collins," *Journal of the Royal Institute of Chemistry* 78 (1954), p. 174.

36. Herbert Yardley, *The American Black Chamber* (Indianapolis: Bobbs-Merrill, 1931). Rpt ed. Annapolis: Naval Institute Press.

37. TNA, Kew, KV 1/74, S. W. Collins to the D.C.C., Report on visit to the U.S.A. as a Special Instructor to the Military Intelligence Branch, May 25 to August 7, 1918.

38. Bruce Norman, *Secret Warfare: The Battle of Codes and Ciphers* (Washington, D.C.: Acropolis, 1973), 93.

39. G. Bruylants, "Récherches expérimentales sur certaines altérations accidentelles ou frauduleuses du papier et de certaines écritures," *Bulletin de l'Académie royale de médicine de Belgique,* 1890, 552–58.

40. J. Thorwald, *Crime and Science* (New York: Harcourt, Brace and World, 1967), 283; W. Jerry Chisum and Brent E. Turvey, *Crime Reconstruction* (Amsterdam: Elsevier/Academic Press, 2011), 33.

41. Edmond Locard, *Manuel de technique policière* (Paris: Payot, 1923), 211.

42. J. Rubner, "Ultraviolette Strahlen und unsichtbare Geheimschrift," *Archiv für Kriminologie* 79, no. 4 (1926), 254–57; F. W. Martin and Konrad Beöthy, "Secret Communications Made Visible," *Lancet,* May 12, 1934, 1006.

43. Kirke Papers 82/28/I, Diary, War Museum, London, June 22, October 11, 1915; Harold Hartley and D. Gabor, "Thomas Ralph Merton, 1888–1969," *Biographical Memoirs of the Fellows of the Royal Society* 16 (Novem-

ber 1970), 421–40, quotation on 425. See also Christopher M. Andrew, *Her Majesty's Secret Service* (New York: Viking, 1986), 149; Keith Jeffery, *Secret History of MI6* (New York: Penguin, 2010).

44. TNA, KV 1/73, Prisoner of War letters, quotation on 30.

45. Ibid., 69.

8

THE UNITED STATES ENTERS THE SECRET-INK WAR

1. U.S. National Archives (NARA), Southeast, Georgia, Passport photo of Bacon and Bertillon file in Atlanta Penitentiary.

2. Sidney Theodore Felstead, *German Scientists at Bay* (New York: Brentano's, 1920); Herbert O. Yardley, *The American Black Chamber* (Indianapolis: Bobbs-Merrill, 1931), rpt. ed. Annapolis: Naval Institute Press; Bacon, quotation: Christopher Andrew, *The Defence of the Realm: The Authorized History of MI5* (London: Allen Lane, 2009), 72–75.

3. Jules Witcover, *Sabotage at Black Tom: Imperial Germany's Secret War in America, 1914–1917* (Chapel Hill, N.C.: Algonquin Books of Chapel Hill, 1989), 99.

4. Felstead, *German Spies at Bay,* 244.

5. Ibid.

6. NARA, RG 65, M-1085, Roll 293, File 2423. A copy of Bacon's narrative is in Maria de Victorica's FBI file. See also Andrew, *Defence of the Realm,* 74.

7. Ibid., 237–38.

8. TNA, Kew, KV 1/74, "The Testing Department M. I. 9. C. Short Report on work done during the War," 137–38, Merton Statement; Harold Hartley and D. Gabor, "Thomas Ralph Merton, 1888–1969," *Biographical Memoirs of the Fellows of the Royal Society* 16 (November 1970), 421–40.

9. NARA, Southeast, Georgia, Federal Penitentiary Files, "George Faux Bacon."

10. Michael Newton, "Gaston Edmond Bayle (1879–1929)," in *The Encyclopedia of Crime Scene Investigation* (New York: Facts on File, 2008), 21–22; "France: Gaston Bayle," *Time,* September 30, 1929.

11. Charles Lucieto, "Invisible Artifice Revealed," *Los Angeles Times,* November 27, 1927; Charles Lucieto, *On Special Missions* (New York: A. L. Burt, 1927).

12. Jacques Boyer, "Reading between the Lines," *Scientific American,* January 1922, 32.

13. Ibid.

14. "France: Gaston Bayle."

15. Felstead, *German Scientists at Bay,* 252–53.

16. TNA, Kew, WO 141/3/4.

17. "Spies Plead Guilty and Keep Secrets," *New York Times,* March 22, 1917.

18. Ibid.

19. NARA, NSA, SRH-030.

20. Ibid. Harvard University Archives, HUD 314.25, HUD 314.50, HUG 300. NARA, RG 457, Box 1124, "Outgrowth of Secret Inks Subsection, MI-8," Lt. Col. A. J. McGrail. See also David Kahn, *The Reader of Gentlemen's Mail: Herbert O. Yardley and the Birth of American Codebreaking* (New Haven: Yale University Press, 2004), 32–33.

21. Harvard University Archives, HUD 314.25, HUD 314.50, HUG 300.

22. David Kahn, *The Codebreakers: The Comprehensive History of Secret Communication from Ancient Times to the Internet,* rev. and updated (New York: Scribner, 1996), 374.

23. H. O. Nolan. *The Production and Detection of Messages Concealed in Writing and Images* (Geneva, Ill.: Riverbank Laboratories, 1918); Kristie Macrakis, *Seduced by Secrets* (New York: Cambridge University Press, 2008), 213.

24. Leo Baeck Institute, New York City, "Maria von Kretschmann Autobiography," April 1919, 48–52.

25. Ibid. Description based on photographs.

26. Ibid., 58.

27. Ibid., 67.

28. Ibid. Search of articles by Mascha Eckmann and MvK (her signature) on Google Books and in newspaper archives.

29. "Maria von Kretschmann Autobiography."

30. Ibid.; Reinhard R. Doerries, "Die Tätigkeit deutscher Agenten in den USA während des Ersten Weltkrieges und ihr Einfluß auf die diplomatischen Beziehungen zwischen Washington und Berlin," in *Diplomaten und Agenten: Nachrichtendienste in der Geschichte der deutsch-amerikanischen Beziehungen,* ed. Reinhard R. Doerries (Heidelberg: Universitätsverlag C. Winter, 2001), 11–52, esp. 42–43.

31. NARA, RG 65, M-1085, Roll 293, File 2423; FBI file, Chronology.

32. Ibid.; the translated letters are in the FBI file; Yardley, *The American Black Chamber,* 93.

33. Yardley, *The American Black Chamber,* 93; "Sight Unseen: Secrets Inks in International Intrigue," *Popular Mechanics,* July 1932, 74–79, esp. 76.

34. NARA, RG 65, M-1085, Roll 293, File 2423; Yardley, *The American Black Chamber,* 108–9.

35. Leo Baeck Institute, New York, Assistant United States Attorney to The Attorney General, April 11, 1918. Curiously, I found this letter in Maria de Victorica's sister's archive with her autobiography, not in the FBI file. The de Victorica FBI file also mentions this confession as key to arresting de Victorica and Wessels.

36. Ibid.; FBI file, Wunnenberg confession.

37. FBI File, Wunnenberg confession. See also Yardley, *The American Black Chamber,* 90–119.

38. FBI File; Jentzer note in FBI file.

39. FBI file; Yardley, *The American Black Chamber,* 117.

40. FBI file, from chronology.

41. Christoph J. Scriba, "Nachruf: Hans Schimank," *Berichte zur Wissenschaftsgeschichte* 4 (1981), 149–53.

42. Letters in FBI file and some partially quoted in Yardley, *The American Black Chamber,* chapter 5.

43. FBI file; "O'Leary Identifies Mme. Victorica at Trial," *New York Times,* July 4, 1918.

44. John Clark Knox, *A Judge Comes of Age* (New York: Scribner, 1941), 115.

45. See, for example, M. H. Mahoney, "Maria Kretschmann de Victorica," in *Women in Espionage: A Biographical Dictionary* (Santa Barbara, Calif.: ABC-CLIO, 1993), 69–71; A. A. Hoehling, *Women Who Spied* (New York: Dodd, Mead, 1967), 75–98.

46. *Current History* (New York: New York Times, 1916–40), 5: 976.

47. Theodore Kytka, "Invisible Photography and Writing, Sympathetic Ink," Classified version of Kytka's 1930 published article, which does not include the toenail engraving, NARA, available online: http://www.archives.gov/press/press-releases/2011/nr11-148.html.

48. See Doerries, "Die Tätigkeit deutscher Agenten in den USA."

49. Quoted in Christopher Andrew, *For the President's Eyes Only: Secret Intelligence and the American Presidency from Washington to Bush* (London: HarperCollins, 1995), 31–46.

9
VISIBLE NAZIS

1. This chapter is based on Kurt Frederick Ludwig's four thousand–page FBI file, British Censorship and MI5 files, Ludwig's Alcatraz prison files, trial testimony, newspaper accounts, and the secondary sources cited in the notes. The account in this paragraph is based on Kurt Ludwig's letters in

the UK National Archives, KV 2/2630–32, FBI files, and Court documents. See also Edward C. Aswell, "The Case of the Ten Nazi Spies," *Harper's,* June 1942, 1; Ladislas Farago, *The Game of the Foxes: The Untold Story of German Espionage in the United States and Great Britain during World War II* (New York: Bantam, 1971), 451; "Girl, 18, Accuses Nazi Spy Suspects," *New York Times,* February 4, 1942; "Girl Spy Unshaken as She Ends Story," *New York Times,* February 7, 1942.

2. Aswell, "Ten Nazi Spies."

3. H. Montgomery Hyde, *Room 3603: The Incredible True Story of Secret Intelligence Operations during World War II* (1962; New York: Lyons, 2001).

4. Ibid., 81.

5. TNA, Kew, DEFE/1/104; William Stevenson, *A Man Called Intrepid* (1976; New York: Lyons, 2000), 172; Hyde, *Room 3603,* 79–80.

6. "Old Bermuda," *Life,* August 18, 1941, 61. *Life* dubbed the women "censorettes." See also *Fodor's 90 Bermuda* (New York: Fodor's, 1990), 45.

7. Interview with Nadya Letteny, June 16, 1970, in Bruce Norman, *Secret Warfare: The Battle of Codes and Ciphers* (Washington, D.C.: Acropolis, 1973), 93.

8. TNA, Kew, DEFE/1/104, DEFE 3/2, World War I secret-ink file.

9. TNA, Kew, DEFE 1/104, "Joe K. German Agent. 25 July 1941: Secret," C. E. Dent, July 29, 1941, "Translation from German" [intercepted letter].

10. Ibid.

11. Stevenson, *A Man Called Intrepid,* 174–75; Hyde, *Room 3603,* 81–83.

12. F. H. Hinsley and S. A. G. Simkins, *British Intelligence in the Second World War,* vol. 4, *Security and Counterintelligence* (New York: Cambridge University Press, 1990), 158.

13. Ludwig's FBI file, SAC Memo, September 12, 1941; numerous laboratory reports in the FBI file; Hyde, *Room 3603,* 84; Michael Sayers and Albert E. Kahn, *Sabotage! The Secret War against America* (New York: Harper and Brothers, 1942), 32.

14. Montgomery Hyde Papers, United Kingdom, Trial Testimony Minutes, Lucy Boehmler; "Man Fleeing U.S. Held as Army Spy," *New York Times,* August 27, 1941.

15. United States Penitentiary, Alcatraz Island, California, Ludwig's Prison File, Special Progress Report April 4, 1952; Conditional Release, monthly reports; Final report.

16. *A Report on the Office of Censorship* (Washington, D.C.: United States Printing Office, 1945), 46–47; Theodore F. Koop, *Weapon of Silence* (Chicago: University of Chicago Press, 1946), 3–15.

17. Thomas Schoonover, *Hitler's Man in Havana: Heinz Lüning and Nazi Espionage in Latin America* (Lexington: University of Kentucky Press, 2008), 16.

18. FBI Headquarters Archive, Heinz Lüning File, 65-44610-164, section 5.

19. Schoonover, *Hitler's Man in Havana*, 56–57.

20. Koop, *Weapon of Silence*, 13–14.

21. National Archives, 457, Interrogation Report of Haehnle. See also Abwehr files, National Archives, T-242. Cf. Farago, *The Game of the Foxes*, 396–97.

22. FBI files on Duquesne Spy Ring; "The World of William Sebold," *Time*, September 22, 1941. See also Alan Hynd, *Passport to Treason: The Inside Story of Spies in America* (New York: Robert M. McBride, 1943).

23. Art Ronnie, *Counterfeit Hero: Fritz Duquesne, Adventurer, and Spy* (Annapolis: Naval Institute Press, 1995), 218.

24. Ibid., 220.

25. FBI files on Duquesne Spy Ring.

26. See David Kahn, *Hitler's Spies: German Military Intelligence in World War II* (1978; New York: Da Capo, 2000), 302.

27. Sayers and Kahn, *Sabotage!* 25–26; Ronnie, *Counterfeit Hero*, 222.

28. FBI files on Duquesne Spy Ring.

29. J. Edgar Hoover, "The Enemy's Masterpiece of Espionage," *Reader's Digest*, April 1946, 1.

30. Ibid.

31. Ronnie, *Counterfeit Hero*, 224–30; Sayers and Kahn, *Sabotage!* 26–27.

32. FBI Story and film *House on 92nd Street;* Ronnie, *Counterfeit Hero*, 235–37.

33. Ronnie, *Counterfeit Hero*, 235–37.

34. "Spies!" *Time*, July 7, 1941.

35. Ibid.

36. Nikolaus Ritter, *Deckname Dr. Rantzau: Die Aufzeichnungen des Nikolaus Ritter, Offizier im Geheimen Nachrichtendienst* (Hamburg: Hoffman and Campe), 1972, 291.

37. Michael Dobbs, *Saboteurs: The Nazi Raid on America* (New York: Knopf, 2004), 9, 53; Eugene Rachlis, *They Came to Kill: The Story of Eight Nazi Saboteurs in America* (New York: Random House, 1961), 21.

38. Rachlis, *They Came to Kill;* Dobbs, *Saboteurs*, 16–17; *Stenographic Transcript of Proceedings before the Military Commission to Try Persons Charged with Offenses against the Law of War and the Articles of War*, Washington, D.C., July 8–31, 1942, 6: 746–64, available at www.soc.umn.edu /~samaha/nazi_saboteurs/nazi06.html.

39. George J. Dasch, *Eight Spies against America* (New York: Robert M. McBride, 1959), 89.

40. *Stenographic Transcript,* line 758; See David Alan Johnson, *Betrayal: The True Story of J. Edgar Hoover and the Nazi Saboteurs Captured during World War II* (New York: Hippocrene, 2007), 127, for picture of handkerchief.

41. *Stenographic Transcript.*

42. Dobbs, *Saboteurs,* 254–57.

10

THE MYSTERY OF THE MICRODOT

1. U.S. National Archives (NARA), RG 65-36994, Dusan M. Popov FBI file.

2. Dusko Popov, *Spy/Counterspy: The Autobiography of Dusko Popov* (Greenwich, Conn: Fawcett, 1975); NARA, RG 65-36994-26, Popov FBI file.

3. US, NARA, RG 65-36994-26, Lanman Report, 9/17/1941.

4. Cf. Ben Macintyre, *Double Cross: The True Story of the D-Day Spies* (London: Bloomsbury, 2012), 39–40, 90–101.

5. "Central Park West Attracts Tenants," *New York Times,* September 17, 1941; Popov, *Spy/Counterspy,* 40.

6. Popov, *Spy/Counterspy,* 40.

7. NARA, RG 65-36994, Connelley to Hoover, August 21, 1941.

8. FDR Library, Hyde Park, Hoover to Edwin M. Watson, September 3, 1941.

9. John F. Bratzel and Leslie B. Rout Jr., "Pearl Harbor, Microdots, and J. Edgar Hoover," *American Historical Review* 87 (December 1982), 1342–51; John F. Bratzel, "Once More: Pearl Harbor, Microdots, and J. Edgar Hoover," *American Historical Review* 88 (October 1983), 953–60; Thomas F. Troy, "The British Assault on J. Edgar Hoover: The Tricycle Case," *Intelligence and Counterintelligence* 3, no. 2 (1989), 169–209; B. Bruce-Briggs, "Another Ride on Tricycle," *Intelligence and National Security* 7, no. 2 (1992), 77–100. Questionnaire available in John Masterman, *The Double-Cross System: The Incredible True Story of How Nazi Spies Were Turned into Double Agents* (1972; rpt. New York: Lyons, 2000), 196–98.

10. Popov, *Spy/Counterspy;* Curt J. Gentry. *J. Edgar Hoover: The Man and the Secrets* (New York: Norton, 1991), 270.

11. Gentry, *J. Edgar Hoover,* 270.

12. TNA, Kew, Popov file.

13. NARA, FBI, RG 65-36994; Russell Miller, *Codename Tricycle: The True Story of the Second World War's Most Extraordinary Double Agent* (London: Pimlico, 2005), 111–13.

14. NARA, FBI, RG 65-6994, Memorandum for Mr. Ladd, December 16, 1941.

15. Ibid.

16. Masterman, *The Double-Cross System,* 79.

17. Joshua Levine, *Operation Fortitude: The Story of the Spy Operation that Saved D-Day* (London: Collins, 2011), 146.

18. U.K. National Archives, WO 208/5175, "Guide to Search for Means of Secret Graphic Communication or Sabotage," June 1944.

19. Nigel West, ed., *The Guy Liddell Diaries,* vol. 1, *1939–1942: MI5's Director of Counter-Espionage in World War I* (New York: Routledge, 2005), 166; TNA, KV4/187.

20. TNA, KV4/187.

21. Frederic Luther, "The Earliest Experiments in Microphotography," *Isis* 41, nos. 3–4 (1950), 277–81; Boris Jardine, "A Collection of John Benjamin Dancer Microphotographs," Explore Whipple Collections, Whipple Museum of the History of Science, University of Cambridge, 2006, available at http://www.hps.cam.ac.uk/whipple/explore/microscopes/microphotographs.

22. John Phin, *The Seven Follies of Science; to Which Is Added a Small Budget of Interesting Paradoxes, Illusions, Marvels, and Popular Fallacies. A Popular Account of the Most Famous Scientific Impossibilities and the Attempts Which Have Been Made to Solve Them. With Numerous Illustrations* (New York: Van Nostrand, 1912), 147.

23. Michael Buckland, *Emanuel Goldberg and His Knowledge Machine* (Westport, Conn.: Libraries Unlimited, 2006), bible quotation on 111, espionage services on 116. See also Michael K. Buckland, "Histories, Heritage, and the Past: The Case of Emanuel Goldberg," in *History and Heritage of Scientific and Technological Information Systems: Proceedings of the 2002 Conference,* ed. W. Boyd Rayward and Mary Ellen Bowden (Medford, N.J.: Published for the American Society of Information Science and Technology and the Chemical Heritage Foundation by Information Today, 2004), 39–45; William White, *The Microdot: History and Application* (Williamstown, N.J.: Phillips Application, 1992), 25–26; William White, "The Microdot: Then and Now," *Intelligence and Counterintelligence,* 3, no. 2 (1989), 249–69.

24. Walter Zapp interview in *The Minox Is My Life,* DVD; autobiographical Notes for Frederic Luther.

25. There was an SD official named Paul Zapp in Dresden in 1943–44 who was responsible for the murder of thousands of Jews during his time in

the Einsatzkommando, but he had no photographic or technical knowledge; Berlin Document Center.

26. NARA, FBI, RG 65-1433, "Johannes Rudolf Christian Zuehlsdorff," Report on De-Briefing. I would like to thank Michael Buckland for sending me a copy of the FBI report. See also Buckland, *Emanuel Goldberg and His Knowledge Machine.* Buckland verified the identity of the new Zapp by contacting staff at the Dresden Technical College department for scientific photography.

27. A Professor Zapp is also listed as head of Amt VI's school of microphotography in a British report given to the United States: "Situation Report No. 11. Amt VI of the RSHA, Gruppe VI F," Counter Intelligence War Room, 30. See NARA, RG 319, IRR case files, RSHA VI; F. Lawrence Malkin, *Krueger's Men: The Secret Nazi Counterfeit Plot and the Prisoners of Block 19* (New York: Little, Brown, 2006); Jacques Delarue, *The Gestapo: A History of Horror* (New York: Skyhorse, 2008); Florian Altenhöner, *Der Mann, der den 2. Weltkrieg begann: Alfred Naujocks: Fälscher, Mörder, Terrorist* (Münster: Prospero Verlag, 2010), 144.

28. J. Edgar Hoover, "The Enemy's Masterpiece of Espionage," *Reader's Digest,* March 1946, 1–6.

29. White, *The Microdot,* 44.

30. FBI, *History of the S.I.S. Division,* 5 vols., May 22, 1947, information on RCA Building facility, 1: 2, declassified and available at www. vault. fbi.gov/special-intelligence-service; Leslie B. Rout Jr. and John B. Bratzel, *The Shadow War: German Espionage and United States Counterespionage in Latin America during World War II* (Frederick, Md.: University Publications of America, 1986), 9–11, 40. See also White, *The Microdot,* 46.

31. William Stevenson, *A Man Called Intrepid* (1976; New York: Lyons, 2000), 372.

32. Rout and Bratzel, *The Shadow War,* 348; David Kahn, *Hitler's Spies* (1978; Cambridge, Mass.: Da Capo, 2000), 322.

33. Kahn, *Hitler's Spies,* 323.

34. FBI, *History of the S.I.S. Division,* 2: 230.

35. On Karl Franz Joachim Rüge see NARA, RG 59, 862.20212/5-2945, 16–19, RG 319; FBI, *German Espionage in Latin America,* 34–35; RG 165, *Mexican Microdot Case #2;* FBI, *History of S.I.S. Division,* 3: 478–80 (here he is called Arnold Karl Franz Joachim Ruge); Rout and Bratzel, *The Shadow War,* 62–65.

36. Rout and Bratzel, *The Shadow War,* 63.

37. NARA, RG 165, FBI, *Mexican Microdot Case,* #2, 41–42.

38. Rout and Bratzel, *The Shadow War,* 75–76.

39. NARA, RG 165, FBI, *Mexican Microdot Case* #1, 59.

40. Stevenson, *A Man Called Intrepid*, 373. Stevenson refers to Rüge as Y2983, whereas the FBI files refer to 2863.

41. Ibid.; NARA RG 319, FBI, German Espionage in Latin America, 36–37; FBI, *History of SIS Division*, 3: 476–81; María Emilia Paz Salinas, *Strategy, Security, and Spies: Mexico and the U.S. as Allies in World War II* (University Park: Pennsylvania State University Press, 1997).

42. FBI, *History of SIS Division*, 3: 480; NARA RG 59, 862.20212/8-1846; Rout and Bratzel, *The Shadow War*, 91.

11
INVISIBLE SPY CATCHERS

1. TNA, KV 2/2632, "German Espionage in the U.S.A. between February 1940 and September 1941," Fol. 116a, p. 13 of report; DEFE 1/103, C. E. Dent, Bermuda, to Watkins-Mence, Fol. 93 d.

2. A. Neuberger, "Charles Enrique Dent," *Biographical Memoirs of Fellows of the Royal Society* 24 (November 1978), 14–31; F. H. Hinsley, *British Intelligence in the Second World War: Its Influence on Strategy and Operations* (New York: Cambridge University Press, 1979), 158.

3. Neuberger, "Charles Enrique Dent."

4. William Stevenson, *A Man Called Intrepid* (1976; New York: Lyons, 2000), 197–98.

5. Ibid.

6. *Life,* May 19, 1941; Milton Bracker, "Bermuda Censors Fill Two Hotels," *New York Times,* May 27, 1941.

7. C. H. "Dick" Ellis, introduction to William Stevenson, *A Man Called Intrepid,* New York: Ballantine Books, 1976, p. xxii. Quotation only in this edition.

8. Christopher M. Andrew, *Her Majesty's Secret Service: The Making of the British Intelligence Community* (New York: Vikings, 1986), 448; Anthony Cave Brown, *"C": The Secret Life of Sir Stewart Menzies* (New York: Macmillan, 1987), 263; H. Montgomery Hyde, *The Quiet Canadian* (London: Hamish Hamilton, 1962), 28. "Eventually to bring the United States into the war" is missing from the American edition of the book, H. Montgomery Hyde, *Room 3603: The Incredible True Story of Secret Intelligence Operations during World War II* (1962; New York: Lyons, 2001), 28. See Thomas E. Mahl, *Desperate Deception: British Covert Operations in the United States, 1939–44* (Washington, D.C.: Potomac, 2000), note 6, for explanation of the change by David Ogilvy. This is why Stephenson was against publication in the United States.

9. Hyde, *Room 3603*, 26.

10. Stevenson, *A Man Called Intrepid*, 172.

11. Hyde, *Room 3603*, 56.

12. G. Wilkinson, "Prof. H. V. A. Briscoe," *Nature* 192 (November 18, 1961), 604; *The Times* (London), September 25, 1961.

13. TNA, Kew, KV 2/1936, Nickolay Hansen file. U.S. National Archives (NARA) OSS files on secret ink contain a note in one of the secret newsletters on quinine ink secreted in a tooth.

14. TNA, Kew, KV 2/1936, Nickolay Hansen file, Professor Briscoe to Mr. Stamp, B.1.B., Minute Sheet, December 2, 1943.

15. *A Report on the Office of Censorship* (Washington, D.C.: United States Printing Office, 1945), 3–4; Mary Knight, "The Secret War of Censors vs. Spies," *Washington Post*, February 3, 1946.

16. *Report on the Office of Censorship;* Knight, "Secret War."

17. "Spy Stories," *Time*, December 28, 1942; NARA, RG 226, 184 A, History of the Technical Operations Division.

18. David Kahn, *The Codebreakers: The Comprehensive History of Secret Communication from Ancient Times to the Internet*, rev. and updated (New York: Scribner, 1967), 515; *A Report on the Office of Censorship*, 43.

19. NARA, RG 226, 184A, 25.

20. Ibid.

21. Harold R. Shaw Report for David Kahn, April 27, 1964, 3, 6, David Kahn Collection, National Security Agency.

22. NARA, RG 226, 184 A, History of the Technical Operations Division, 1–3, 51.

23. Ibid.

24. A *Report on the Office of Censorship*, 10, 43; Theodore F. Koop, *Weapon of Silence* (Chicago: University of Chicago Press, 1946), 77; Kahn, *The Codebreakers*, 517–18.

25. NARA, RG 226, folder 1, The History of the Technical Operations Division, Exhibit "c." organizational chart. The women listed were Miss Jean Brengle, Mrs. Irene Link, Mrs. Julia F. J. Woods, and Mrs. Ethel T. Pierce.

26. Harold R. Shaw Report, April 27, 1964, David Kahn Collection.

27. NARA, RG 226, 184A, Conferences, Summaries of Miami Conference, 1943; Summary of Charles de Graz Lecture.

28. NARA, RG 226, 184A, Conferences.

29. Ibid.

30. Ibid.

31. NARA, RG 226, History of the Technical Operations Division, 17.

32. Irvin Stewart, *Organizing Scientific Research for War: The Adminis-*

trative History of the Office of Scientific Research and Development (Boston: Little, Brown, 1948), esp. chapter 2, "National Defense Research Committee," 7–34, and chapter 4, "NDRC of OSRD—The Committee," 52–78; William Albert Noyes and Ralph Connor, *Chemistry: A History of the Chemistry Components of the National Defense Research Committee, 1940–1946* (Boston: Little, Brown, 1948); Vannevar Bush, *Pieces of the Action* (New York: Morrow, 1970).

33. NARA, RG 226, 184A, 3, The History of the Technical Operations Division, Volume VIII, Laboratory Research and Recommendations, 2–3; Letter Byron Price to Vannevar Bush, January 27, 1944; Bush to Price, March 15, 1944.

34. NARA, RG 226, 184A, 3, The History of the Technical Operations Division, Volume VIII, Laboratory Research and Recommendations, 2–3; Letter Byron Price to Vannevar Bush, January 27, 1944; Bush to Price, March 15, 1944.

35. Linus Pauling Papers, Oregon State University Special Collections, video clip and transcript available at http://osulibrary.oregonstate.edu/spe cialcollections/coll/pauling/war/audio/hager2.006.3-writing.html.

36. Pauling Papers, Report #4, Outline of Method 1., February 20, 1945, George W. Wright, Frank Lanni, William Eberhardt, Linus Pauling, Official Investigator.

37. Sanborn Brown's work is documented in the minutes to the monthly OSRD meetings, 1944–45, NARA RG 226/Entry 184 A/Box 3, Minutes of meetings June 1944–September 1945. See also "The Use of X-Ray for the Detection of Microdots" in the same file. When David Kahn interviewed him in 1963, Sanborn did not provide details about his wartime work on microdots; Kahn Interview with Sanborn Brown, 1963, David Kahn Papers, National Security Agency Library.

38. NARA, RG 226/Entry 184 A/Box 1, Report on history of TOD; RG 226/184A/Box 2, Digest of the Miami Conference, S. W. Collins on "Future Research: What Science Can and Cannot Do," 56–57.

39. NARA, RG 226/Entry 184 A/Box 3, Laboratory Research file, 8–9 of report on visits to various institutions, MIT, May 1, 1943; RG 226/27, "Massachusetts Institute of Technology OEMsr-1403 Report. Pats, Infra-Red SW, and Radioactive SW," December 31, 1945.

40. NARA, RG 226/211/29, "The Use of Compounds of Europium in SW," John W. Irvine Jr., Sanborn C. Brown, Robley D. Evans, May 24, 1945.

41. TNA, Kew, DEFE 1/103, French Report on German Apparatus, original and translation.

42. TNA, Kew, DEFE1/ 103, Note by Professor Briscoe on the French

Report Concerning the Apparatus Abandoned by the Germans in Paris, February 27, 1945.

43. Colonel Shaw report, April 27, 1964, David Kahn Papers; NARA, RG 226/184A/3, OSRD Minutes for Fifteenth Meeting of the Defense Committee, Harvard University, August 3, 1945; RG 226/27, "Massachusetts Institute of Technology OEMsr-1403 Report, Pats, Infra-Red SW, and Radioactive SW," December 31, 1945.

44. Richard Breitman et al., *U.S. Intelligence and the Nazis* (Cambridge: Cambridge University Press, 2005), 93.

45. Ibid.; Ladislas Farago. *Burn after Reading: The Espionage History of World War II* (Annapolis: Naval Institute Press, 2003), 11.

46. David Kahn, *Hitler's Spies: German Military Intelligence in World War II* (1978; New York: Da Capo, 2000), 279.

47. David Kahn Papers, National Security Agency Library, Interview Albert Müller with David Kahn, April 1970.

48. Ibid.

49. NARA RG 457, Box 202, Folder VO-63, Dr. Hans Otto Haehnle Interrogation Report, British Origin, May 12–15, 1946, Bad Salzufen, 10. Also available in NARA RG 226, Box 353.

50. Ibid.

51. Ibid.

52. TNA, Kew, WO 204/12461, Extract from CI News Sheet No. 11.

53. Ibid., G. E. Kirk to Captain H. Grotrian, April 17, 1944.

12
OUT IN THE COLD

1. BStU (Stasi Archive Files), GH 21/84, vol. 1, fol. 130, Wolfgang Reif file, including interrogation minutes, the Military Court Papers, minutes, and transcripts of the proceedings. There is also a film of some of the proceedings; JHS, no. 313/89, Wolfgang Jatzlau, "Examination of the Historical Development of Department M in the 70s," March 3, 1989, fol. 17–18; weather report from www.tutiempo.net/en/Climate/BERLIN_DAHLEM/12-1981/103810 .htm.

2. Stasi file, vol. 1, fol. 5, 128.

3. Ibid.; Stasi file, fol. 322, Jatzlau; Helmut Wagner, *Schöne Grüße aus Pullach: Operationen des BND gegen die DDR* (Berlin: Das Neue Berlin, 2001), 127–28.

4. Stasi files, vol. 1, fol. 5, 128, fol. 322; Wagner, *Schöne Grüße aus Pullach*.

5. Stasi file, vol. 8, fol. 219.

6. Kristie Macrakis, *Seduced by Secrets: Inside the Stasi's Spy-Tech World* (New York: Cambridge University Press, 2008), 185, based on BStU, GVS, MfS 008-Nr. 1002/68, 9. Richtlinie 2/68 des Ministers für Staatssicherheit.

7. Robert Wallace and H. Keith Melton with Henry R. Schlesinger, *Spycraft: The Secret History of the CIA's Spytechs from Communism to Al-Qaeda* (New York: Dutton, 2008), 392–93.

8. BStU, MfS, OTS files for aspirin secret ink. For the alleged East German foreign intelligence molar concealment see the spy exhibit at Times Square and accompanying catalogue, *Spy: The Secret World of Espionage* (n.p., n.d.), 27.

9. Wallace and Melton with Schlesinger, *Spycraft*, 58–60.

10. Ibid., 20–21; Macrakis, *Seduced by Secrets,* 145.

11. Wallace and Melton with Schlesinger, *Spycraft,* 60.

12. William L. Cassidy, "Sympathetic Inks: A Study of Secret Writing," *Interservice Journal* 1, no. 1 (n.d.), 45–49.

13. Benjamin B. Fischer, "The Central Intelligence Agency's Office of Technical Service, 1951–2001," brochure, 20.

14. Kristie Macrakis, Elizabeth K. Bell, Dale L. Perry, and Ryan D. Sweeder, "Invisible Ink Revealed: Concept, Context, and Chemical Principles of 'Cold War' Writing," *Journal of Chemical Education* 89 (2012), 529–32; Kristie Macrakis, *Seduced by Secrets,* 205–10.

15. A. K. Gupta, A. Lal, et al., "Electrostatic Detection of Secret Writing," *Forensic Science International* 41 (1989), 17–23; E'lyn Bryan, "Questioned Document Examination," *Evidence Technology Magazine* 8, no. 3 (2010), 20–24.

16. BStU, MfS, JHS, MF. GVS 112/84; Renate Murk, "Increasing the Quality of an Operational-technical Investigative Method to Detect Secret Writing Traces and Secret Writing on Operationally Relevant Letters," thesis, MfS Law School, October 4, 1983.

17. Richard Tomlinson, *The Big Breach: From Top Secret to Maximum Security* (Edinburgh: Cutting Edge, 2001), 68.

18. Ibid.

19. BStU, HA IX, no. 961, bd. 2, fol. 46.

20. David Kahn, *The Codebreakers: The Comprehensive History of Secret Communication from Ancient Times to the Internet,* rev. and updated (New York: Scribner, 1996), 515, 524.

21. BStU, OTS, 2038, fol. 57. JHS, no. 313/89; fol. 45, Jatzlau, "Examination of the Historical Development."

22. Wallace and Melton with Schlesinger, *Spycraft,* 64.

23. *Final Report of the Select Committee to Study Governmental Opera-*

tions with Respect to Intelligence Activities, United States Senate: Together with Additional, Supplemental, and Separate Views (Washington, D.C.: Government Printing Office, 1976), Domestic CIA and FBI Mail Opening Programs, book 3, April 23, 1976, 1.

24. Memorandum from A. H. Belmont to D. E. Moore, March 10 1961, reprinted in *Hearings before the Select Committee to Study Governmental Operations with Respect to Intelligence Activities, United States Senate, Ninety-Fourth Congress, First Session* (Washington, D.C.: Government Printing Office, 1976), vol. 4, Mail Opening, exhibit 22, 244 (commonly known as the Church Committee Report).

25. Ibid.; Memorandum from Chief, CI/Project to DC/CI, August 30, 1971, exhibit 5, 199.

26. "Special Report Interagency Committee on Intelligence (ad hoc)," Chairman, J. Edgar Hoover, June 1970, 31; Exhibit 11, p. 218 of *Hearings before the Select Committee to Study Governmental Operations with Respect to Intelligence Activities.*

27. Wolfgang Jatzlau, "Examination of the Historical Development," fol. 17–18.

28. *Final Report,* 53.

29. "Thompson Denies Spy Charges and Says He's '100% American,'" *New York Times,* January 9, 1965.

30. Robert Glenn Thompson with Harold H. Martin, "I Spied for the Russians," *Saturday Evening Post,* part 1, May 22, 1965, 23–29, esp. 26; part 2, June 5, 1965, 38–49.

31. Ibid., 1: 26.

32. Ibid., 2: 39–40.

33. Ibid.

34. Ibid., 2: 40.

35. Ibid.

36. Nigel West, *Historical Dictionary of Cold War Intelligence* (Lanham, Md.: Scarecrow, 2007), 340–42.

37. Ibid.

38. BStU, MfS, HA IX, no. 961, bd. 2, fols. 47–51.

39. Wallace and Melton with Schlesinger, *Spycraft,* 430–31.

40. *Arabesque,* Universal Studios, 1966.

41. *You Only Live Twice,* United Artists, 1967.

42. Jim Stockdale and Sybil Stockdale, *In Love and War,* rev. and updated (Annapolis: Naval Institute Press, 1990), 194–97; Lorraine Adams, "Perot's Interim Partner Spent 7½ Years as POW," *Seattle Times,* March 31, 1992, www.community.seattletimes.nwsource.com/archive.

43. Stockdale and Stockdale, *In Love and War,* 194–95.

44. Wallace and Melton with Schlesinger, *Spycraft,* 300–303.

45. FBI File on Aryan Brotherhood, available in the FBI Vault; Christopher Goffard, "Invisible Ink Got Gang's Deadly Note Past Guards," *Los Angeles Times,* June 27, 2006, http://articles.latimes.com/2006/jun/27/local/me-code27.

46. Samuel Rubin, *The Secret Science of Covert Inks* (Port Townsend, Wash.: Loompanics, 1987), 27.

47. Mano Alexandra, *Secret Love Letters and the Legend of the Lovers from Prague* (San Francisco: Chronicle, 2006).

48. Ibid.

49. Jean Radakovich to the author, May 2012; see www.radakovich.org/documentary.html.

13
HIDING IN PORN SITES

1. Jack Kelley, "Terror Groups Hide Behind Web Encryption," *USA Today,* February 5, 2001; Niels Provos and Peter Honeyman, "Detecting Steganographic Content on the Internet," *CITI Technical Report 01-11,* August 31, 2001, quotation from abstract; Niels Provos and Peter Honeyman, "Hide and Seek: An Introduction to Steganography," *IEEE Security and Privacy,* May–June 2003, 32–44.

2. Yassin Musharbash, "In ihren eigenen Worten," *Die Zeit,* March 15, 2012, www.diezeit.de/2012/12/Al-Kaida-Deutschland/kompettansicht; Nic Robertson, Paul Cruickshank, and Tim Lister, "Documents Reveal Al-Qaeda's Plans for Seizing Cruise Ships, Carnage in Europe," CNN, May 1, 2012.

3. Eric Cole, *Hiding in Plain Sight: Steganography and the Art of Covert Communication* (Indianapolis: Wiley, 2003), 5.

4. Gina Kolata, "Veiled Messages of Terror May Lurk in Cyberspace," *New York Times,* October 30, 2001.

5. Neil F. Johnson information page, John and Johnson Technology Consultants, LLC, www.jjtc.com/neil; Kolata, "Veiled Messages."

6. Kolata, "Veiled Messages."

7. Ibid.

8. Gregory Kipper, *Investigator's Guide to Steganography* (Boca Raton, Fla.: Auerbach, 2004), 6–7; G. S. Simmons, "The Prisoner's Problem and the Subliminal Channel," *CRYPTO 83, Advances in Cryptology,* August 22–24, 1984, 51–67.

9. Federal Bureau of Investigation (FBI), "Sealed Complaint. United

States of America vs. Christopher Metsos, Richard Murphy, Cynthia Murphy, Donald Howard Heathfield, Tracey Lee Ann Foley, Michael Zottoli, Patricia Mills, Juan Lazaro," June 25, 2010, 5. A second FBI affidavit against Anna Chapman and Mikhail Semenko is dated June 27, 2010.

10. Toby Harnden, "Richard and Cynthia Murphy: 'Suburbia's Spies Next Door,'" *Telegraph,* July 4, 2010; David Montgomery, "Arrest of Alleged Spies Draws Attention to Long Obscure Field of Steganography," *Washington Post,* June 30, 2010.

11. FBI, "Sealed Complaint," 23–24.

12. Ibid., 9–10.

13. Ibid.

14. Ibid., 15.

15. Ibid. For quotation see "Spy Swap Russian: I'll Go Home But Only if It's Safe," *London Evening Standard,* August 18, 2010.

16. FBI, "Sealed Complaint."

17. Henry Fountain, "Hiding Secret Messages within Human Code," *New York Times,* June 22, 1999.

18. Catherine Taylor Clelland, Viviana Risca, and Carter Bancroft, "Hiding Messages in DNA Microdots," *Nature* 399 (June 10, 1999), 533–34.

19. Ibid.

20. D. G. Gibson et al., "Creation of a Bacterial Cell Controlled by a Chemically Synthesized Genome," *Science* 329 (2010), 52–56.

21. Manuel A. Palacios et al., "InfoBiology by Printed Array of Microorganism Colonies for Timed and On-demand Release of Messages," *Proceedings of the National Academy of Sciences* 108, no. 40 (2011), 16510–14; Ed Yong, "Bacteria Encode Secret Messages," *Nature,* September 26, 2011, http://www.nature.com/news/2011/110926/full/news.2011.557.html.

22. Palacios et al., "InfoBiology by Printed Array."

EPILOGUE

1. "CIA De-Classifies Oldest Documents in U.S. Historical Collection," April 19, 2011, https://www.cia.gov/news-information/press-releases-statements/press-release-2011/cia-declassifies-oldest-documents-in-u.s.-government-collection.html. There was a blizzard of news articles after the CIA's press release. See, for example, Peter Finn, "CIA Recipe for Invisible Ink among Newly Released WWI-Era Documents," *Washington Post,* April 19, 2011. http://articles.washingtonpost.com/2011-04-19/world/35230951_1_secret-ink-invisible-ink-documents; "U.S. District Court Rules That World War I German Invisible Ink Formulas Must Still

Remain Hidden from the Public," James Madison Project, February 4, 2002, www.jamesmadisonproject.org/press.php?press_id=19. The released file contains sixteen pages if one does not delete the duplicate pages. Even though the National Archives was the de jure defendant, the CIA was the de facto defendant.

2. NARA, RG457, 190/22/5/1.

3. Ibid.

4. James Bamford, "Inside the Matrix," *Wired*, April 2012, 78–84, 122–24.

APPENDIX

1. TNA Ho 144/20116.

2. Adapted from Johann Nikolaus Martius, *Unterricht in der natürlichen Magie, oder zu allerhand belustigenden und nützlichen Kunststücken*, totally rev. Johann Christian Wiegleb (Berlin: Friedrich Nicolai, 1779), 195–98.

PRIMARY SOURCES

Archival

Federal Bureau of Investigation: Kurt Ludwig File, Fritz Duquesne Spy Ring

Federal Commissioner for the Records of the State Security Service of the former German Democratic Republic (BStU)

France: Caen Archives, National Archives, Academy of Sciences Archive

German Federal Archives, Berlin

German Federal Military Archives, Freiburg

Harvard University Archives, Theodore W. Richards, Arthur B. Lamb

National Cryptological Museum, David Kahn Papers

Oregon State University Special Collections, Linus Pauling Archive

United Kingdom: Montgomery Hyde Papers

United Kingdom: The National Archives (TNA), Kew, Richmond, Surrey (formerly PRO); Primary Record Groups included KV (MI5), DEFE (Ministry of Defense), WO (War Office)

United States: U.S. National Archives and Records Administration (NARA), RG 65 (FBI), Maria de Victorica, Dusan M. Popov; RG 319. U.S. Censorship Records

United States National Archives, Atlanta branch, San Bruno, California, branch, New York State branch

Digital Primary Sources

British History Online (www.british-history.ac.ak)

Early English Books Online (EEBO)

Eighteenth Century Collections Online (ECCO)

Google Books

Google NGram Viewer (www.books.google.com/ngram)

Royal Society Archive
State Papers Online, 1509–1603

Printed Sources and Rare Book Libraries

Boston Athenaeum Library
Chemical Heritage Foundation, Rare Books Library
Houghton Library, Harvard University
Huntington Library
Newberry Library, Chicago
The University of Chicago Rare Books Room
Wellcome Institute History of Medicine Library

CREDITS

ACKNOWLEDGMENTS

I started this book during my sabbatical year, 2007–8, as a visiting scholar at the Department of History of Science at Harvard University and a member of the Boston Athenaeum. I would like to thank Anne Harrington, the chairwoman, and Everett Mendelsohn for their warm welcome and support at Harvard. Thanks also to Mark Kornbluh for facilitating the sabbatical.

In the course of researching the book I visited a variety of archives and rare book rooms in the United States and Europe to conduct research, and I'm grateful for the help of archivists and librarians, starting, of course, with those at Harvard's own Houghton Library. Internal grants from my universities helped support this research. In England, the staff at the National Archives in Kew and the Wellcome Institute were very helpful as I sought sources in person and, later, electronically. In Germany, the Bundesarchiv in Berlin and its military branch in Freiburg also aided in my search, along with the BStU, the Federal Authority for the Stasi Archives. In the United States, the National Archives in College Park, Maryland, was a rich source of material, along with its subsidiaries in Atlanta, New York, and San Bruno, California. Thanks also to the FBI FOIA division for sending me materials related to my topic. The interlibrary loan staff at Georgia Tech was extraordinarily helpful in finding and delivering books related to secret ink and the sociopolitical context of the times I was studying. Several students—Zach Meisel, Ella Smith, and Taylor Prichard—were delightful research assistants and readers of drafts. Warm thanks to the Huntington Library staff for sending me photographs. Three digital sources were crucial to accelerating research for this project: Early English Books Online (EEBO), Eighteenth Century Collections Online (ECCO), and Google Books. I would also like to thank the Chemical Heritage Foundation for providing me with a grant to visit and do research at the center and present a talk.

Thanks to Melissa Chinchillo, my incomparable agent, for her enthusiastic support of the project, and to Vadim Staklo for commissioning it for Yale University Press. After Laura Davulis took over as editor at Yale she kindly read the whole manuscript and provided helpful comments as I revised it.

I wrote the majority of the book in 2012. I am particularly grateful to members of my writer's group in Atlanta—the Frantic Muse—for their support as I raced to complete the manuscript in time for the December 2012 publisher's deadline. Andrew Sutter and Cliff Bostock started the group, and Pam Perry read and provided insightful comments on most of what I submitted. I am particularly grateful to Andrew Sutter for introducing me to Jason Lye, a color chemist who read several chapters and contributed to the appendix. He proved to be a delightful partner as we conducted invisible-ink experiments in his kitchen.

I am grateful to a number of friends and colleagues who agreed to read part or all of the manuscript. Dr. Lily Macrakis read the whole manuscript during her stay in Atlanta and provided encouragement and a general reader's reactions. I was very happy when my old mentor and friend Michael Kater, one of the world's leading experts on Nazi Germany, agreed to read the three Nazi Germany chapters and the introduction. I am very appreciative of his unfailing support over the years and very helpful comments. My colleague at Georgia Tech Jonathan Schneer read several chapters and provided insightful comments. I am also grateful to Ron Bayor and Steve Usselman for their support of my research. Several experts read drafts of earlier incarnations of the ancient and early modern chapters. In particular, I am grateful to Glen Bowersock and Katherine Park. The anonymous reviewers for Yale also provided helpful feedback. Finally, this book is dedicated to David Kahn, author of the monumental *Codebreakers* and the world's leading authority on cryptology. Not only did he provide encouragement when I toyed with the idea of writing about this elusive topic, but he led me the right way for initial sources. If it weren't for his support and belief in the topic, I might never have written the book.

INDEX

Abracadabra, 109
Abwehr, 175, 178, 180, 183–86,
    188–89, 192, 194, 199, 201–6,
    208, 212, 214, 216–17, 230, 243,
    245–50, 258, 338; secret commu-
    nication department, IG, 212,
    246–47; in South America, 214;
    stations in Berlin, Hamburg, and
    Cologne, 216
Accademia dei Secreti, 23
Acid, 58
AEG, 209
Aeneas, Tacticus, 10–11
Africa, 144
Agentenfunk (Afu), 215. *See also*
    Radio
Age of Wonder, 83
Agfa, 217
Alberti, Leon Battista, 27–28
Alcatraz Island Prison, 181
Alchemist, 71–72, 81, 104, 118
Alchemy, 16, 48, 71–72, 118
Alexandra, Mano, 283
Alonso, Manuel, 178
Al-Qaeda, xii, 212, 285
Alum, 13–14, 24–26, 32, 37, 114,
    306, 311
Aluminum chloride, 150

Aluminum potassium sulfate. *See*
    Alum
American Black Chamber, The, 146
American intelligence, 84
American Journal of Police Science,
    298
American Revolution, 84–85, 98
American Revolutionary War,
    83–103, 240
Ammann-Brass, Hans, 211–12
Ammonia, 196, 308–9
Ammonium chloride, 57
Ammonium hydroxide, 196
Ammonium vanadate, 250
André, John, 100
Angels, 64, 69
Anhydrous, 80, 123
Antwerp, 128, 131, 163
Apis, 185. *See also* aspirin
Apothecaries, 110–11
Aqua Fortis, 82
Aqua Regis, 77, 81. *See also* Royal
    acid
Arabesque, 277
Arabs, 14
Arden, John, 43
Argentina, 214–15
Argyrol, 157